RAND Health NATIONAL DEFENSE RESEAR

Final Evaluation Report for Uniformed Services Family Health Plan Continuous Open Enrollment Demonstration

Michael Schoenbaum

Katherine Harris

Gary Cecchine

Melissa Bradley

Ana Suarez

Terri Tanielian

C. Ross Anthony

Prepared for the Office of the Secretary of Defense
Approved for public release; distribution unlimited

This report was conducted jointly by RAND Health's Center for Military Health Policy Research and the Forces and Resources Policy Center of the National Defense Research Institute, a federally-funded research and development center sponsored by the Office of the Secretary of Defense, the Joint Staff, the unified commands and the defense agencies. It was performed under TRICARE contract DASW01-95-C-0059, "TRICARE Senior Supplement and Uniformed Services Family Health Plan Demonstration Evaluations."

ISBN: 0-8330-3004-3

RAND is a nonprofit institution that helps improve policy and decisionmaking through research and analysis. RAND® is a registered trademark. RAND's publications do not necessarily reflect the opinions or policies of its research sponsors.

Published 2001 by RAND
1700 Main Street, P.O. Box 2138, Santa Monica, CA 90407-2138
1200 South Hayes Street, Arlington, VA 22202-5050
201 North Craig Street, Suite 102, Pittsburgh, PA 15213-1516
RAND URL: http://www.rand.org/
To order RAND documents or to obtain additional information, contact Distribution Services: Telephone: (310) 451-7002; Fax: (310) 451-6915; Internet: order@rand.org

Preface

The National Defense Authorization Act for Fiscal Year (FY) 2000 established the authority for a demonstration program under which eligible beneficiaries were to be permitted to enroll in the Uniformed Services Family Health Plan (USFHP) at any time. Congress directed that the Department of Defense (DoD) test the feasibility and advisability of providing "continuous open enrollment" for a two-year period at a minimum of two of the seven USFHP sites; submit a report to Congress by March 15, 2001, evaluating the benefits and costs of the program; and make a recommendation concerning whether to authorize continuous open enrollment at all USFHP sites on a permanent basis.

RAND performed the evaluation for the TRICARE Management Activity of the Office of the Secretary of Defense for Health Affairs as part of a project entitled "TRICARE Senior Supplement and Uniformed Services Family Health Plan Demonstration Evaluations." The work was carried out jointly by RAND Health's Center for Military Health Policy Research and the Forces and Resources Policy Center of the National Defense Research Institute. The latter is a federally funded research and development center sponsored by the Office of the Secretary of Defense, the Joint Staff, the unified commands, and the defense agencies.

The report should be of interest to members of Congress, DoD, the seven USFHP programs, and military retirees and their dependents.

Contents

Tables

Summary

The National Defense Authorization Act for Fiscal Year (FY) 2000 established the authority for a demonstration program under which eligible beneficiaries were to be permitted to enroll in the Uniformed Services Family Health Plan (USFHP) at any time. We will refer to this policy as "continuous open enrollment." Congress directed that the Department of Defense (DoD) test the feasibility and advisability of providing continuous open enrollment for a two-year period at a minimum of two (out of seven) USFHP sites; submit a report to Congress by March 15, 2001, evaluating the benefits and costs of the program; and make a recommendation concerning whether to authorize continuous open enrollment at all USFHP sites on a permanent basis.

Background

The USFHP sites are designated providers of the TRICARE Prime program (TRICARE Prime is one of the three components of TRICARE, the health insurance program for current and former military personnel and their dependents). Except for active duty personnel, all military beneficiaries living in the seven USFHP service areas are eligible to enroll in and receive care from USFHP. USFHP is currently the only TRICARE product available to Medicare-eligible military beneficiaries.

Military retirees and their dependents who live in a USFHP service area have had the option of directly enrolling in USFHP during a designated 30-day period in the spring of each year; the exact period varies by site. In contrast, eligible military beneficiaries (i.e., those under age 65) can enroll in TRICARE Prime at any time. Enrollment in USFHP is for a one-year period, during which enrollees are locked out of receiving medical care at Military Treatment Facilities (MTFs) or from other TRICARE providers. Medicare-eligible beneficiaries are also instructed not to use their Medicare benefit while enrolled in USFHP, on penalty of being disenrolled.

Covered benefits for all enrollees are, at minimum, the same as those covered by TRICARE Prime. For Medicare-eligible enrollees, USFHP thus covers some services that Medicare does not, particularly prescription drugs and preventive services. USFHP receives capitated payments from DoD for enrolled military beneficiaries; i.e., the Uniformed Services Treatment Facility (USTF) gets a fixed, prospective payment per beneficiary from DoD to cover the beneficiary's medical care costs for a specific period of time. Capitation payments are based on beneficiaries' age; sex; and, in some sites, on health status as measured by prior health service use.

Demonstration

The TRICARE Management Activity randomly selected three sites to participate in the demonstration: PacMed, Bayley Seton, and Brighton Marine. The demonstration was to begin on October 1, 1999, and continue for two years. Continuous open enrollment began on January 1, 2000, at Bayley Seton; on March 1, 2000, at PacMed; and on April 1, 2000, at Brighton Marine.

Methods

To assess the likely impact of continuous open enrollment on beneficiaries, DoD, and other stakeholders, we conducted interviews with USFHP and DoD staff and other key participants involved in rate-setting negotiations; site visits at each demonstration site; focus groups with enrolled and eligible non-enrolled beneficiaries at each demonstration site; and analyses of administrative data on enrollment patterns under the demonstration.

Findings

Consequences and Costs for Beneficiaries

- The continuous open enrollment policy had no apparent drawbacks for beneficiaries.
- To date, we have found no evidence that the continuous open enrollment policy increased enrollment in USFHP. However, this finding should not be considered definitive, because of the design and duration of the demonstration.
- For beneficiaries under age 65, we found that a *de facto* policy of continuous open enrollment already exists, because beneficiaries may enroll in TRICARE Prime at any time and then transfer to USFHP at any time. Providing direct access to USFHP via continuous open enrollment would reduce the administrative burden on beneficiaries wishing to enroll.

- We found evidence that, in the absence of continuous open enrollment, some Medicare-eligible beneficiaries spend periods without supplemental insurance (i.e., Medigap) or purchase relatively expensive private Medigap coverage while waiting to enroll in USFHP. The continuous open enrollment policy improves access to USFHP and makes such situations less likely. This benefit will increase in the likely event that Medicare HMOs discontinue operations or restrict new enrollment in USFHP service areas.
- The relative value of continuous open enrollment for Medicare-eligible beneficiaries is likely to decline with the introduction of the new health insurance benefits under the FY2001 Defense Authorization Act.

Consequences and Costs for DoD

- Data and design limitations, along with other contemporaneous network expansions at several of the demonstration sites, prevented us from making reliable and valid predictions of the effects of the continuous open enrollment policy on overall enrollment and on the distribution of health risks in USFHP.
- DoD expressed concern that continuous open enrollment would promote adverse selection. We expect that the financial risk to DoD of such enrollment is likely to be minor, mainly because the current system of calculating capitation rates places most risk from adverse selection on USFHP. (We were unable to conduct an empirical evaluation of adverse selection during the demonstration because of data limitations.)
- The continuous open enrollment policy may ultimately produce a net increase in USFHP enrollment. In particular, the relative generosity of USFHP benefits, combined with more-comprehensive marketing efforts made feasible by continuous open enrollment, may increase enrollment in areas where awareness of the program has been low.
- The new health insurance benefits for Medicare-eligible military beneficiaries are likely to reduce substantially the relative value of USFHP for that population, and thereby the net cost of continuous open enrollment to DoD. The magnitude of these effects will depend on the details of how the new benefits are structured.

Recommendations

We recommend that continuous open enrollment be extended to all USFHP sites on a permanent basis, conditional on this policy being acceptable to the USFHP programs themselves. Short of this result, we recommend that the enrollment policies of USFHP match those of TRICARE Prime for beneficiaries who are eligible for both programs.

Acknowledgments

We are grateful to Susan Hosek, Jeffrey Wasserman, and Dave Kennel for their comments on an earlier draft, Sydne Newberry for editorial advice, and Kristine Leiphart for research assistance.

1. Introduction

The National Defense Authorization Act for Fiscal Year (FY) 2000 established the authority for a demonstration program under which covered beneficiaries were to be permitted to enroll in the Uniformed Services Family Health Plan (USFHP) at any time. We refer to this as "continuous open enrollment."[1] Congress directed that the Department of Defense (DoD) test the feasibility and advisability of providing continuous open enrollment for a two-year period at a minimum of two (out of seven) USFHP sites; submit a report to Congress by March 15, 2001, evaluating the benefits and costs of the program; and make a recommendation concerning whether to authorize continuous open enrollment at all USFHP sites on a permanent basis. The demonstration was scheduled to begin on October 1, 1999, and end on September 30, 2001.

The FY2001 Defense Authorization Act substantially expanded the DoD health insurance benefits available to Medicare-eligible beneficiaries, including coverage for prescription drugs via the National Mail Order Pharmacy and TRICARE eligibility for all military retirees and their dependents regardless of age. The act significantly changes the policy context in which this demonstration takes place. Although little detail was available by the deadline for this report regarding how the new benefits will be implemented, we examined the text of the legislation and conducted informal conversations with TRICARE Management Activity (TMA) staff about the likely nature of the new benefits. This information guided our interpretation of the findings of this evaluation, where appropriate.

Section 2 of this report provides background on the USFHP program and describes the continuous open enrollment demonstration. Section 3 describes the goals of this evaluation, and Section 4 describes the methods we used to meet these goals. Section 5 describes our evaluation results. Section 6 presents our conclusions and recommendations, as well as the limitations of the study.

---

[1]In this report, "beneficiaries" refers to individuals who are eligible to participate in TRICARE or USFHP by virtue of serving or having served in the United States military, and their dependents.

2. Background

USFHP Program Overview

Under the Omnibus Budget Reconciliation Act of 1981, 10 of the 35 Public Health Service hospitals operating at that time were transferred to local, nonprofit private ownership and became Uniformed Services Treatment Facilities (USTFs) under contract to DoD. Of these 10 facilities, one was subsequently sold and is no longer a USTF, and three merged into one. There are currently seven independent programs, each serving a specific geographic service area:

- **USFHP—Johns Hopkins Medical Services Corporation**, based in Baltimore, MD, and serving the District of Columbia and parts of Maryland
- **USFHP—Brighton Marine Health Center**, based in Boston, MA, and serving parts of Massachusetts, Rhode Island, Connecticut, and New Hampshire
- **USFHP—CHRISTUS Health**, based in Houston, TX, and serving parts of Texas and Louisiana
- **USFHP—Sisters of Charity Medical Center at Bayley Seton**, based in Staten Island, NY, and serving parts of New York and all of New Jersey
- **USFHP—PacMed Clinics**, based in Seattle, WA, and serving parts of Washington
- **USFHP—Martin's Point Health Care**, based in Portland, ME, and serving Maine
- **USFHP—Fairview Health System**, based in Cleveland, OH, and serving parts of Ohio.

The USFHP sites are designated providers of the TRICARE Prime program.

TRICARE is the health care program for active-duty members of the uniformed services and their dependents, retired military personnel and their dependents, and surviving dependents of military personnel. For eligible beneficiaries, the TRICARE program offers a triple-option health care plan: (1) TRICARE Prime, which integrates military treatment facilities (MTFs) and civilian health care into a single, managed delivery system; (2) TRICARE Standard, a fee-for-service option in which beneficiaries can choose to receive care from most civilian physicians; and (3) TRICARE Extra, which is similar to TRICARE Standard but offers discounts to patients when they use TRICARE network providers. To participate in TRICARE Prime, beneficiaries enroll for a one-year period. There is no enrollment requirement for TRICARE Standard or Extra.[1]

With the exception of active-duty personnel, military beneficiaries living in the geographic service area of one of the USFHP sites are eligible to enroll in and receive care from USFHP. The eligible population thus includes military retirees and their covered dependents, and dependents of active-duty personnel, regardless of age. UFHP is the only TRICARE option available to Medicare-eligible retirees.

Each program provides a full range of inpatient and outpatient medical services as well as prescription drugs. Covered benefits for all enrollees are, at a minimum, the same as those covered by TRICARE Prime. For Medicare-eligible enrollees, USFHP thus covers some services that Medicare does not, particularly prescription drugs and preventive services.

USFHP sites are organized as health maintenance organizations (HMOs), using managed-care practices commonly associated with HMOs (e.g., medical care is coordinated by primary care providers, with additional utilization review for inpatient care). USFHP receives capitated payments from DoD for enrolled military beneficiaries. Under capitated payment, the USTF receives a fixed, prospective payment per beneficiary from DoD to cover the beneficiary's medical care costs for a specific period of time; this payment does not change regardless of what actual costs turn out to be. Historically, capitation payments were based solely on beneficiaries' age and sex. Following the FY1998 National Defense Authorization Act, capitation payments are also now required to take beneficiaries' health status into account. DoD payments to USFHP are discussed in additional detail below.

Medical care under USFHP is provided by a mix of clinicians, some of whom are USFHP employees and others of whom are independent clinicians with USFHP contracts. Similarly, care is provided in a mix of USFHP-owned facilities (i.e., hospitals and clinics) and independent facilities with USFHP contracts. Prescription drugs are dispensed via a mix of USFHP-owned pharmacies and independent pharmacies with USFHP contracts. In each case, the mix varies by site.

USFHP enrollment precludes beneficiaries from receiving any medical care from MTFs, other TRICARE Prime providers, or under the TRICARE Standard or

[1]Source: http://www.tricare.osd.mil/tricare/news/faqs.htm.

Extra options. In addition, Medicare-eligible enrollees are instructed that any use of their Medicare benefits during their USFHP enrollment is grounds for terminating their enrollment in USFHP. Beneficiaries whose enrollment is terminated are prohibited from re-enrolling for one year.

In addition to serving military beneficiaries, some USFHP sites serve other populations. These activities are outside the scope of the demonstration and our evaluation, and we do not discuss them further in this report.

USFHP Payment Process

Overview

USFHP was specifically designated in the DoD budget through FY1996. Since then, the USFHP program has been covered by the general DoD health budget. The contracts of each USFHP site stipulate that DoD must approve all enrollment in USFHP to ensure that expenditures do not exceed funding levels. Overall enrollment in USFHP, across all sites, cannot grow by more than 10 percent per year, excluding dependents of active-duty personnel. However, the USFHP program has never hit this cap.[2]

USFHP capitation payments are negotiated bilaterally with DoD through five-year, sole-source contracts called "individual participation agreements."

Rate Setting

With the exception of Johns Hopkins,[3] capitation payments for the six other USFHP sites are calculated in a roughly similar fashion, with separate rates for enrollees under and over age 65.[4] Payments for enrollees under age 65 are based on average military health system (MHS) costs per beneficiary, within gender and age cells. These costs are calculated as the sum of third-party claims, direct care, administration, medical education, and resource sharing costs generated by MHS users in each cell, divided by the number of "reliants" in the cell, where reliance on the MHS is measured through surveys of the MHS-eligible

[2]Norbert Meister, personal communication, August 28, 2000.

[3]The Johns Hopkins contract differs because it is located in Baltimore, Maryland, an all-payer-rate state.

[4]Rate-setting formulas are specified in Section C10 in each of the seven contracts.

population.[5] Payments are calculated for 28 cells based on seven age categories and gender, and they are adjusted for geographic variations in costs according to a formula based on the regional cost of the Federal Employees Health Benefit Program, the Medicare Prospective Payment System, and the Resource Based Relative Value Scale.

The payments for Medicare-eligible enrollees are divided into two parts. Payments for Medicare-covered services are based on the Adjusted Average Per Capita Cost (AAPCC) rate for the appropriate age, gender, and county cell. The AAPCC rates are inflated (roughly equal to the AAPCC/.95) to reflect changes over time in the method used by the Health Care Financing Administration (HCFA) to calculate AAPCCs and the fact that USFHP sites do not actually have Medicare Risk Contracts. AAPCC-based payments make up roughly 85 percent of total payments for Medicare-eligible enrollees. The other 15 percent represent the cost of the pharmaceutical benefit and preventive services not otherwise covered by Medicare and thus are "new" government expenditures.

Although each of the six contracts (i.e., excluding Johns Hopkins) follows the same basic structure, individual sites have negotiated special provisions to account for factors, such as regional geography or use patterns, that affect the cost of providing care at the local level.

Risk Adjustment

Current contracts with Martin's Point, CHRISTUS Health, Brighton Marine, and PacMed have health status adjuster clauses that are specific to the Medicare-eligible enrollees. These sites have been able to prove to DoD's satisfaction that they are being adversely selected based on ambulatory care group (ACG) methods,[6] which compare the distribution of diagnosed morbidity of USFHP enrollees and a 5 percent sample of Medicare beneficiaries, respectively.[7] Johns Hopkins has not negotiated risk-adjusted payments, and Bayley Seton was not able to supply the necessary data because of disruptions suffered in the wake of the insolvency of its provider network administrator.

Methods used for risk-adjusting USFHP payments are unrelated to those currently being phased in by Medicare for paying managed-care plans, although

---

[5]Several study participants expressed concern that the methodology used to measure reliance overstated the number of "reliants" in a service area, with the effect of reducing the resulting capitation payment.

[6]The ACG risk adjustment system was developed by Jonathan Weiner and colleagues at Johns Hopkins University; see http://acg.jhsph.edu/.

[7]Adverse selection in this case means that USFHP enrollees are sicker, on average, than their eligible counterparts who do not enroll in USFHP and on whom the capitation rates are based.

the USFHP contracts specify that DoD will take risk adjustment into account in calculating USFHP rates when HCFA uses it to calculate the AAPCCs. However, the choice of risk-adjustment methodology may become an issue in future negotiations.

Enrollment Policies

Current Policy

Dependents of active-duty personnel can enroll in USFHP at any time; they are thus outside the scope of the demonstration and our evaluation, and we do not discuss them further in this report.

Military retirees and their dependents who live in a USFHP service area have had the option of enrolling in USFHP during a designated 30-day period in the spring of each year; the exact period varies by site. Enrollment is for a 12-month period. There are several exceptions to these policies:

- Newly retired beneficiaries and their dependents who live in a USFHP service area may enroll in USFHP at the time of retirement.
- Eligible beneficiaries moving from an area without a USFHP program to a USFHP service area may enroll in USFHP when they move.
- Enrollees in one USFHP site who move to the service area of another USFHP site may transfer enrollment between sites, and non-Medicare-eligible USFHP enrollees who move to an area with no USFHP program may transfer to TRICARE Prime. This is referred to as "portability."
- USFHP enrollees who move outside the service area may transfer to TRICARE Prime at any time. If they travel temporarily (e.g., if they are snowbirds), they may transfer back to USFHP one time during the year.
- Eligible beneficiaries who are enrolled in TRICARE Prime, whether through a managed-care support contractor or through an MTF, may currently transfer to USFHP at any time. This is referred to as "transferability." Such a transfer is permitted once per year.

The last exception is particularly relevant for this evaluation. In particular, current DoD policies permit eligible retired beneficiaries and their dependents to enroll in TRICARE Prime at any time, and then to transfer to USFHP. This creates a *de facto*, indirect policy of continuous open enrollment in USFHP for beneficiaries who are eligible for TRICARE Prime.

Continuous Open Enrollment Demonstration

Starting in 1998, the USFHP sites independently and collectively asked TMA to permit retired beneficiaries and their dependents to enroll in USFHP at any time, as the dependents of active-duty personnel were permitted to do.[8] The stated motivation for this request was to make USFHP more competitive with TRICARE Prime. In particular, this would affect retired beneficiaries and their dependents who were permitted to enroll in TRICARE Prime at any time but could enroll in USFHP only during the designated one-month open enrollment period (at the time of this request, the policy of "transferability" had not yet been instituted).

USFHP personnel expressed concern that the existing policy of annual open enrollment put USFHP at a competitive disadvantage with respect to TRICARE Prime, for several reasons. First, newly eligible beneficiaries who delayed enrolling because they were unfamiliar with USFHP would need to wait a year for the next opportunity to enroll. Second, like USFHP, TRICARE Prime requires a 12-month enrollment commitment, and, in general, the benefit year for TRICARE Prime enrollees would not align with USFHP's enrollment cycle. As a result, TRICARE Prime enrollees who wished to enroll in USFHP would face a period between the end of a TRICARE Prime benefit year and the next USFHP enrollment period during which they would not be enrolled in either plan (although they could receive care through TRICARE Standard or Extra during this period).

USFHP's 1998 request for continuous open enrollment represented a reversal of its previous position, which USFHP personnel acknowledged in their correspondence. In particular, in the course of prior contract negotiations, USFHP had rejected such a policy when it was proposed by DoD.

Later in 1998, TMA rejected USFHP's request to change the enrollment policy.[9] The stated motivation for this denial was concern that continuous open enrollment would expose DoD to increased costs because of adverse selection (i.e., an enrollment pattern in which beneficiaries who enrolled due to the continuous open enrollment policy were sicker, on average, than those who would otherwise have enrolled), especially by Medicare-eligible beneficiaries

---

[8]For instance, in letters from Stephen Weiner (on behalf of Brighton Marine) to Diane Harler (Contracting Officer, Defense Supply Service), dated June 8, 1998; from Tita Montero (USFHP Director, PacMed) to JoAnna Black (TRICARE Northwest), dated June 12, 1998; David Howes (Program Director, Martin's Point) to Earl Hanson (Contracting Officer Representative, TMA), dated June 12, 1998; and Mark Skulnick (VP of Finance, Sisters of Charity) to Diane Harler, dated June 15, 1999.

[9]For instance, in letters from Diane Harler to David Howes, dated October 9, 1998; and from Diane Harler to Mark Skulnick, dated October 9, 1998.

who would not otherwise be eligible for DoD health insurance benefits. DoD personnel expressed concern that adverse selection would increase DoD costs because of the recently instituted policy of accounting for beneficiaries' health status. DoD personnel pointed out that DoD's original request for continuous open enrollment had been made prior to the adoption of this risk-adjustment policy and that USFHP's rejection of continuous open enrollment at that time was based on concern about adverse selection.

Congress subsequently passed the National Defense Authorization Act for FY2000, which authorized a demonstration of continuous open enrollment in at least two USFHP sites. USFHP personnel requested that all seven sites be permitted to implement continuous open enrollment under the demonstration, with the stated motivation being that differences across sites would make findings difficult to generalize to other sites.[10] Ultimately, TMA randomly selected three sites to participate in the demonstration: PacMed, Bayley Seton, and Brighton Marine.

The demonstration was to begin on October 1, 1999. In practice, continuous open enrollment began on January 1, 2000, at Bayley Seton;[11] on March 1, 2000, at PacMed;[12] and on April 1, 2000, at Brighton Marine.[13] In the absence of the demonstration, these sites would have conducted the following open enrollment periods:

- **Bayley Seton:** July, effective August 1
- **Brighton Marine:** April, effective June 1
- **PacMed:** February, effective April 1.

During the remaining 11 months of the year, waiting lists would have been maintained with the names of eligible beneficiaries expressing interest in the program.

[10]See correspondence from Stephen Weiner (on behalf of the seven USFHP sites) to Charles Abell (Senate Armed Services Committee), dated June 30, 1999; and from the seven USFHP Program Directors to James Sears (Executive Director, TMA), dated October 4, 1999.

[11]Eric Feldman, personal communication, October 4, 2000.

[12]Tita Montero, personal communication, September 27, 2000.

[13]David Chicoine, personal communication, September 27, 2000.

3. Purpose of Evaluation

According to the National Defense Authorization Act for FY2000, the evaluation of the demonstration "shall include, at a minimum, an evaluation of the benefits of the open enrollment opportunity to covered beneficiaries and a recommendation concerning whether to authorize open enrollments in the managed-care plans of designated providers permanently." Given this broad mandate, we focused this report on the following issues:

- Are there benefits to continuous open enrollment for beneficiaries? Are there drawbacks?
- Does continuous open enrollment change overall enrollment in USFHP or the distribution of demographic and health characteristics among enrollees?
- Would continuous open enrollment have cost implications for beneficiaries, USFHP, the Department of Defense, or the federal government?
- What are the pros and cons of making the demonstration national and permanent? How might the program differ if it were made permanent?

As mentioned above, we interpret our findings in the context of the new health benefits included in the FY2001 Defense Authorization Act, as appropriate.

4. Methods

Secondary Data on USFHP Sites

As a starting point, we reviewed several types of materials on the demonstration sites and, as appropriate, on other USFHP sites. First, we obtained program descriptions and sample enrollment materials from each of the three demonstration sites. We also reviewed the Internet web sites for the demonstration sites (available via links from www.usfhp.org) and for the other four USFHP sites. We used these materials to get a basic descriptive understanding of the USFHP programs before our site visits and interviews with USFHP and DoD personnel.

Second, we obtained the actual contracts between DoD and the three demonstration sites. Although these contracts provided valuable information on the financial relationship between DoD and the USFHP, we were particularly interested in assessing the distribution of financial risks to the respective parties from possible adverse selection that were due to the demonstration (i.e., whether the continuous open enrollment policy led particularly sick beneficiaries to enroll who otherwise would have enrolled later or not at all). We discuss this issue in greater detail below. The contracts appeared to be complete, save for the methodology for calculating capitation rates. We received separate memoranda from Kennell and Associates (actuarial consultant to TMA) describing the methodology used to calculate capitation rates, including information on the mechanisms used for adjusting capitation rates to reflect enrollees' health status.

Next, we obtained information on Medicare HMO plans (Medicare Choice) in each of the three demonstration sites. We used these materials to assess the range of insurance choices available to Medicare-eligible military beneficiaries in the demonstration sites. No comparable, comprehensive information was available on the insurance choices that might have been available to military beneficiaries below age 65. However, we obtained information on the availability of MTF care and TRICARE Prime, at MTFs and via managed-care support contractors.

Interviews

We conducted interviews with the following key individuals:

- Earl Hanson, contracting officer representative, TMA
- David Kennell, Kennell and Associates (actuarial consultant to TMA)
- Tita Montero, director, PacMed
- Eric Feldman, acting director, Bayley Seton
- David Chicoine, director, Brighton Marine
- Norbert Meister, USFHP program coordinator.

We used these interviews to learn about the history of the USFHP program generally and of the enrollment policy in particular, the nature of the relationships between DoD and USFHP, the pros and cons of the continuous open enrollment policy from the perspective of the various interview subjects, and any other issues the interview subjects thought would help us conduct the evaluation.

In addition, we attended the quarterly meeting of the USFHP site directors on September 21, 2000, in Washington, DC. The site directors of all USFHP sites were present, as were additional staff from each site. Since available resources did not permit us to conduct site visits at the non-demonstration sites, this meeting was our main opportunity to speak with key personnel from these sites.

Site Visits

We made visits to each of the three demonstration sites: Bayley Seton (August 17, 2000), Brighton Marine (August 18, 2000), and PacMed (August 22, 2000). At each visit, we interviewed the USFHP program director and other staff responsible for the administration of the program and toured selected USFHP clinics. We did not seek to speak with patients or other USFHP enrollees on these site visits.

We used these visits to learn about the history of the USFHP program at that site, the current and historical organization of the site's delivery system, descriptive information about the site's providers and enrollees, the nature of the relationships between the site and USFHP, the pros and cons of the continuous open enrollment policy from the perspective of the site visit participants, and any other issues the interview subjects thought would help us conduct the evaluation.

In addition, because eligible beneficiaries who choose to enroll in USFHP are effectively agreeing to receive all their medical care from USFHP (since it functions as an HMO, with a closed panel of providers), we were interested in visiting selected clinics to get a qualitative picture of the environment where enrollees receive care. In practice, we visited the clinic closest to the administrative headquarters of each site, which we recognize might not be representative of USFHP clinics overall.

Enrollment Data

To examine patterns of enrollment, we obtained data on USFHP enrollees from the Iowa Foundation for Medical Care (a DoD contractor responsible for maintaining these data and for collecting and processing USFHP claims and encounter data) and on eligible beneficiaries from the Defense Enrollment Eligibility Reporting System (DEERS).

At a conceptual level, enrollment due to the demonstration would consist of beneficiaries enrolling who would not have enrolled under the other policy or beneficiaries enrolling earlier or later than they would have otherwise. Beneficiaries making such enrollment decisions cannot be individually identified, because each beneficiary faces only one of the two enrollment policies at any point in time. They could be identified implicitly in the context of a randomized control trial, e.g., if half the eligible beneficiaries in a given site had been randomly assigned the opportunity to enroll on a continuous basis, while the other half could enroll only during one month per year.

In the absence of such randomization, we could compare eligible and newly enrolled beneficiaries in demonstration and non-demonstration sites, respectively. However, such analyses would need to be interpreted cautiously, because the various USFHP sites appeared to differ in ways that we could not account for in our evaluation, including organizational structure and the non-DoD insurance alternatives available to DoD beneficiaries. Alternatively, we could compare populations within a site before and after the onset of the demonstration. This would also have a number of limitations, particularly if—as we report below—other changes in the USFHP programs or the broader health insurance environment were taking place concurrently with the demonstration. In practice, we conducted both types of analyses to approximate the ideal analyses, and we discuss their limitations in greater detail below.

Focus Groups

We conducted focus groups in each of the three demonstration sites: Bayley Seton (November 28–29, 2000), Brighton Marine (December 11–12, 2000), and PacMed (December 5–6, 2000). Participants included USFHP enrollees and eligible beneficiaries who were not currently enrolled. The focus group protocols were designed to assess beneficiaries' opinions about and satisfaction with USFHP or their chosen alternative health plans, factors affecting their decision to enroll or not enroll in USFHP, and their opinions and experiences with the demonstration.

In addition to the focus groups, we discussed with TMA whether to conduct a survey of eligible and enrolled beneficiaries to collect quantitative data on the issues addressed qualitatively in the focus groups. Together with TMA, we concluded that the project timeline and resources did not permit us to develop, field, and analyze a survey that would add sufficient value beyond the other evaluation activities.

Site Selection

We selected sites in the respective USFHP service areas based on the density of recently enrolled beneficiaries within a 15-mile radius, the availability of adequate facilities, and the ease with which participants could reach the chosen location in rush hour traffic. To reflect both historical and more recent enrollment patterns, we conducted one set of focus groups near Boston, MA (part of the original Brighton Marine service area), and another set near Camden, NJ (part of the recent expansion of the Bayley Seton service area). We conducted the final set of focus groups between Seattle and Everett, WA, in the PacMed service area.

Recruitment

We held separate sessions for beneficiaries over and under age 65 because of differences in the set of health plan alternatives available to these respective groups. In each of the three sites, we combined enrollees and eligible non-enrollees in sessions for those under age 65. Because of DoD and USFHP's expressed concern about the possibility that continuous enrollment would especially affect Medicare-eligible beneficiaries, we conducted separate sessions for enrollees and eligibles age 65 and over at two of the three sites (because of the small number of recently enrolled beneficiaries age 65 and over at our Brighton Marine site, we pooled all Medicare-eligible participants in one group there).

Our goal was to confirm 12 participants for each focus group, with the expectation that 8 to 10 would actually attend. Lists of potential participants were randomly selected for each session. Data came from the Iowa Foundation and DEERS, as described above. For each focus group, sponsors (military retirees themselves) were twice as likely to be sampled as adult dependents (i.e., spouses; we excluded children from this analysis), to ensure that no more than nine dependents would be included in any given focus group. Table 4.1 lists the distribution of focus group participants.

Table 4.1

Focus Group Participants

USFHP Site	Group Composition	Date	Enrollees	Eligibles
Bayley Seton	<65, mixed	11.28.2000	8	3
	65+, eligible	11.29.2000	1	6
	65+, enrolled	11.29.2000	9	0
PacMed	<65, mixed	11.5.2000	6	2
	65+, eligible	11.5.2000	7*	5
	65+, enrolled	12.6.2000	13	0
Brighton Marine	<65, mixed	12.11.2000	7	4
	65+, mixed	12.12.2000	6	3
TOTAL			57	23

* In this case, we neglected to exclude enrollees when sampling eligibles.

Lists for the Bayley Seton and PacMed focus groups included between 30 and 40 potential participants each. Given the small number of beneficiaries in the Boston area who had enrolled since the onset of the demonstration, we included all such enrollees as potential participants, along with two random samples of 20 names of non-enrollees. For the PacMed and Brighton Marine groups, we subsequently had to sample additional names in order to reach the target number of confirmations (for Brighton Marine, we ultimately had to add several beneficiaries who had enrolled in USFHP prior to the demonstration period, although recently).

Potential focus group participants received a recruitment letter from RAND accompanied by an endorsement letter from either the site director (enrollees) or study sponsor (eligibles). (Sample letters are included as Appendices E and F to this report.) This letter was then followed by a phone call from RAND to confirm participation. Finally, we sent confirmation letters to beneficiaries who agreed to participate. (Sample letters are included as Appendices G and H.)

Although we reached the recruitment goals for the number of participants in each focus group, in two cases the groups did not have the intended composition. Specifically, one participant in the "65+, eligible" group for Bayley Seton had enrolled in USFHP too recently for the enrollment to be reflected in our data. More significantly, 7 of 12 participants in the "65+, eligible" group for PacMed were actually enrolled in USFHP, because we accidentally excluded only enrollees who had enrolled since the onset of the demonstration from being sampled for that group.

Discussion Structure and Content

The focus group discussions were conducted according to a set interview guide (included as Appendix I to this report). The recruitment letters asked participants to expect the focus groups to last 90 minutes. In general, discussions were completed in the allotted time. Each focus group began with introductions and a description of the purpose of the discussion. Confidentiality issues were discussed and participants were reminded of the voluntary nature of their participation.

In order to determine what factors influenced participants' decisions to enroll in USFHP or alternative health plans, participants were provided with preprinted index cards listing factors that might influence individuals' health plan choices. The cards were presented in random order and participants were first asked to select out those factors that had any influence in their decision. Any factors that were not considered were laid aside. The preprinted factors included the following:

- Choice of doctors
- Cost
- Coverage
- Desire for military-sponsored health care
- Drug benefit
- Location
- Only option
- Quality
- Recommended by health care provider
- Recommended by friends or family.

In addition, blank cards were provided so participants could add factors important to them. Following this, participants were asked to rank the factors they had selected by relative importance in their decision process, with the most important reason ranked as 1. The cards were then collected, in the order in which participants had sorted them; we subsequently coded the cards with a ranking of 1 to 10 for each participant, 1 indicating the highest ranking. Factors not included by participants were considered to be of the lowest importance and were assigned a rank of 11. Limited descriptive analyses were performed for the participants as a whole and for subgroups according to Medicare eligibility, USFHP enrollment status, and location. Given the limited sample size and the focus group design, we view these analyses—along with the rest of the focus group results—as qualitative data informing our conclusions.

Claims and Encounter Data

We obtained data on USFHP medical claims and encounters from the Iowa Foundation with the intention of using them to help assess possible adverse selection that was due to the continuous open enrollment policy. However as described in Appendix A, we concluded that the data currently available would not support valid and generalizable analyses. We therefore examined adverse selection solely on a conceptual level.

5. Findings

USFHP Program Operations

Prior to conducting focus groups and analyzing data on enrollment, we examined the organization and operation of the demonstration sites using published materials, site visits and interviews with staff, and program materials provided by USFHP and TMA. The purpose of this examination was to assess factors in the organizational structure and geographic location of the USFHP facilities as well as information on competing managed-care organizations in the areas of interest that might influence enrollment. Our main substantive finding from the site visits was that the USFHP programs changed contemporaneously with the continuous open enrollment demonstration, and are likely to continue to change, in ways that plausibly affect the impact of the open enrollment policy. In particular, two of the three demonstration sites (Brighton Marine and Bayley Seton) have recently expanded their provider networks and service areas; indeed, these sites reported that new enrollment, including during the demonstration period, was occurring disproportionately in the expansion areas. We include a more complete description by site in Appendix B.

Health Insurance Market Environment

The USFHP program draws enrollees from two distinct populations with two distinct sets of health insurance alternatives, summarized in Table 5.1. Beneficiaries age 65 and over can choose between USFHP, Medicare managed-care plans, fee-for-service Medicare, fee-for-service Medicare with a private supplemental insurance ("Medigap") plan, and post-retirement health insurance coverage offered by a former civilian employer. Beneficiaries under 65 choose from USFHP, other TRICARE plans, or health plans sponsored by civilian employers.

The availability, cost, and generosity of each of these alternative sources of care vary from site to site and affect the impact of the demonstration on overall enrollment and mix of health risks in USFHP. Nonetheless, our site visits revealed two trends that were consistent across sites. First, site visit participants at Bayley Seton and Brighton Marine felt that interest in USFHP among those

Table 5.1

Health Insurance Options Available to USFHP Eligibles

Beneficiary Group	Health Insurance Options
Non–Medicare Eligible (under age 65)	USFHP TRICARE Prime TRICARE Standard/Extra Health plans provided by current and former civilian employers
Medicare Eligible (age 65 and over)	USFHP Private Medicare Supplemental Insurance ("Medigap") Medicare HMOs Health plans provided by current and former civilian employers

under 65 was driven by the lack of well-developed and popular TRICARE networks, so that USFHP served as the main viable TRICARE Prime option in the area. Participants in both sites felt that this competitive advantage was likely to erode over time as Managed Care Support Contractors gained more experience in the area. PacMed differed, because alternative TRICARE providers, particularly the Madigan MTF, were relatively well established in the Seattle market.

Second, participants felt that interest in USFHP among those 65 and older was driven by the relative generosity of the prescription drug benefit compared with that available through local Medicare HMOs and Medigap plans. In each demonstration site, a number of Medicare HMOs had recently announced that they were closing, which could cause some military beneficiaries to shift to USFHP. More importantly, however, site visit participants in all sites indicated that the relative attractiveness of USFHP would fall or even disappear with the anticipated (and now actual) enactment of the new health insurance benefits for military retirees in the FY2001 Defense Authorization Act, particularly the drug benefit.

Appendix C contains additional information on the Medicare HMO alternatives in the three demonstration sites.

Continuous Open Enrollment and Marketing Activities

Site visit discussions about marketing activities focused on efforts to attract military retirees, the group for whom restricted annual open enrollment had previously applied. Participants at Brighton Marine and Bayley Seton agreed that the market for retirees residing in close proximity to the former public health hospital facilities was stable, saturated, and satisfied; participants at PacMed, where the service area has been largely unchanged in recent years, expressed a similar view. Participants shared the view that a strong sense of entitlement had gone a long way in assuring that all eligibles who wanted to enroll had done so and that disenrollment rarely occurred for reasons other than death.

The lack of potential new members in the traditional service areas prompted both Brighton Marine and Bayley Seton to expand their provider networks in an effort to attract new enrollees. Site visit participants shared the view that the annual open enrollment requirement hampered efforts to attract members in the new service areas, many of whom were unaware of their eligibility for the program.

Site visit participants at the three demonstration sites and representatives from the other USFHP sites reported that the daunting logistics of mounting marketing campaigns in geographically diffuse areas in the short period of time had substantially increased their marketing costs, and that continuous open enrollment would reduce costs by allowing them to use resources more rationally over longer periods of time (e.g., by not having to hire temporary workers). In the participants' view, continuous open enrollment would end frustration at having to turn away people who inquired about the program after the enrollment deadline had passed.

However, site visit participants in all demonstration sites also expressed concern that continuous open enrollment would remove some of the sense of urgency that helped to assure that potential enrollees would submit application forms in a timely manner. Without the pressure of limited enrollment, they said, potential new members were more likely to procrastinate.

Participants also indicated that retirees under age 65 were entering the program though the "back door" anyway, by enrolling in TRICARE and transferring to USFHP in an unrestricted fashion. However, participants at Bayley Seton and Brighton Marine did not report actually counseling potential members to pursue this strategy, but participants at PacMed indicated that they did actively recommend this path between the introduction of "transferability" and the onset of the demonstration.

Participants at all three demonstration sites warned that many enrollees and eligible beneficiaries would probably not know that their ability to enroll in USFHP had ever been restricted. They suggested that 40–60 individuals formerly on the waiting list might possibly be aware of the policy change. They were also keenly aware of the difficulty of distinguishing those who enrolled because of the new policy from those who would have enrolled under the old system.

Appendix D includes additional detail by site on marketing strategies.

Enrollment—Focus Group Findings

In this subsection, we report on the results of the focus groups. Our main findings include a generally high level of satisfaction with the USFHP program among enrollees and a relatively low level of awareness, not only about the continuous open enrollment demonstration but also about the program itself, among eligible beneficiaries. No beneficiaries believed that continuous open enrollment would be less desirable for them than the current policy of annual enrollment; indeed, many felt the new policy would be advantageous, and some reported negative consequences of the annual enrollment policy.

Factors Influencing Enrollment Decisions

In aggregate, cost and coverage were consistently ranked as the most important factors influencing participants' decisions to join USFHP or another health plan (See Table 5.2). These factors were followed in aggregate by a set of factors that included quality, location, choice of doctors, drug benefit, and a desire for military-sponsored health care. "Only option" and recommendations by friends, family, and health care providers were generally ranked as least important. The blank cards were rarely used and most often restated the preprinted factors. For example, one Medicare-eligible participant enrolled in USFHP added "promised lifetime health care," which was determined during discussion to relate closely to "desire for military health care."

The ranking of all factors differed little between Medicare-eligible and non-Medicare-eligible participants, and Medicare eligibility appeared to have little effect on participants' ranking of factors that affected their health plan choices. Regardless of Medicare eligibility, USFHP enrollees placed substantially more importance on a desire for military-sponsored health care than did non-enrollees for the selection of their respective health plans. Cost was also consistently ranked as a more important factor among USFHP enrollees than among non-enrollees. Non-USFHP enrollees placed more importance on coverage and choice of doctors than did enrollees, and non-enrollees also cited recommendations by health care providers as a more important factor than did enrollees. Finally, enrollees consistently ranked drug coverage as more important than did non-enrollees, regardless of their Medicare eligibility.

Satisfaction with Current Coverage

Each of the focus groups included substantial discussions regarding customer satisfaction with USFHP. In general, few differences were identified between

Table 5.2

Selected Comments from Focus Group Participants

Decision Factors*	Enrollees	Non-Enrollees
Cost	An important factor in using MHS benefits prior to joining USFHP. The cost of the program has increased over time, but remains relatively low. USFHP low cost compared to COBRA and other insurance options. No other realistic alternative; the only "real" choice, despite co-pays. Former plan raised premiums. Never get a bill.	Not a concern in current plan because it doesn't charge patients.
Coverage	Coverage has declined over the years.	
Quality	USFHP has good doctors.	
Location	Stayed in the area because of USFHP. The most important factor.	Brighton Marine facility perceived as inconvenient.
Choice of doctors	The inability to stay with previous doctor was a drawback of USFHP. A very important factor. Preferred doctor on the USFHP list.	Important to maintain relationship with current physician. A perceived drawback of USFHP. A confusing aspect of USFHP (PPO vs. HMO lists). Not enough specialists in USFHP.
Drug benefit	Not much of a benefit because you still have to pay. MHS drug benefit more attractive overall than USFHP. Better than Aetna. However, it is best to get drugs through the Army.	Drug costs can be reduced by using mail order plans. An important factor because drugs at MTFs are often not available and you must wait in long lines to get them.
Desire for military-sponsored health care	An important factor in choosing Brighton Marine. You know you are in good hands. A promise of lifetime health care made to us.	Desirable but not readily accessible. Few USFHP physicians in the immediate area. MHS not reliable source of coverage, better to choose a more expensive plan. More concerned about access; waits too long at MTF.

Table 5.2 (continued)

Decision Factors*	Enrollees	Non-Enrollees
		Experienced troubles using TRICARE Prime that are still not resolved.
Recommended by health care provider		Enrolled in plan recommended by current doctor.
Recommended by friends or family		The most important factor in choosing a health plan.
Only option		

* List appears in overall order of importance based on results of card sort activity.

Medicare-eligible and non-Medicare-eligible groups. Both USFHP enrollees and non-enrollees cited a high level of satisfaction with their present health plan, which was in accordance with their reasons for selecting it.

Interestingly, while enrollees most often cited cost as an important factor in choosing a health plan, many of the participants expressed displeasure at having to pay any premiums or co-payments for what they perceived as an entitlement promised to them for free in return for military service. Although drug benefits did not rank as high as cost and coverage when participants gave their reasons for selecting a health plan, the USFHP drug benefit was frequently cited as beneficial or even essential to the financial stability of some enrollees, particularly those who are Medicare-eligible and those with chronic illnesses. The quality of care provided at USFHP facilities was frequently described by enrollees as very high, but some enrollees in Bayley Seton and Brighton Marine also cited difficulties with access (particularly dental services and optometry, which are covered in limited amounts), high turnover in program physicians, and lack of coverage for specialized medical supplies as factors that created dissatisfaction (Table 5.3).

Reasons for Not Enrolling in USFHP

Considering the commonly high level of satisfaction among USFHP enrollees, participants were asked to comment on why more military retirees do not enroll in USFHP (see Table 5.4). Responses fell into three general categories. First, participants in all three demonstration areas stated simply that not enough information about USFHP was disseminated to the eligible population. In short, it was the opinion of the participants that not many people know of the program and that more retirees would enroll if more information were consistently available. A substantial portion of participants indicated that they enrolled after hearing about the program from friends, relatives, or organizations (e.g., Retired

Table 5.3

Selected Focus Group Comments Regarding Satisfaction with USFHP

Site	Comment
Bayley Seton	USFHP has pretty much the same doctors as other HMOs or health plans.
	Tests were avoided and spouse sent home from the hospital still sick.
	Patient needed a hip replacement, but other treatments were tried before approval for surgery was given.
	USFHP takes too long to reimburse doctors—credit records could be harmed.
	Satisfaction was expressed with services and choice of doctors.
	The drug benefit is very good.
	Administrative problems were encountered with paperwork; no 1-800 number was available to connect directly to Bayley Seton.
Brighton Marine	Plan lacks out-of-area coverage.
	Beneficiaries can't use MTF out of area.
	Turnover rate among doctors is high.
	Dental and eye benefits are problems.
PacMed	Non-emergency appointments are delayed.
	Coverage area expanded without expanding physician pool.
	Beneficiaries were concerned about continuity and recruitment of primary care physicians.
	Twenty-four-hour telephone access to nurses is an important benefit.
	Differences between PacMed and USFHP are not well understood.
	Beneficiaries were very satisfied.
	Beneficiaries were concerned about impact of recent legislation on the plan.

Officers' Association). Non-enrollees generally agreed, and at least half of these participants asked where they could get more information about USFHP.

Second, participants in all the demonstration areas and particularly in the Seattle area indicated that some retirees depend on health care benefits from a second career (e.g., Boeing is a large employer in Seattle). Some participants indicated that comparing these benefits with USFHP was difficult. Several participants voiced a perception that USFHP, as a government-sponsored program, was likely to be discontinued, so alternative health plans—if affordable—provide better security.

Third, participants commonly relayed anecdotes about counter-marketing or misinformation from other TRICARE officials. For example, USFHP was recommended to one participant's son, whose retirement from the military is pending. When the son called an MHS representative to ask about USFHP, he was informed that the representative was permitted to discuss only TRICARE, not USFHP. Similar stories were offered by other participants, who were informed by TRICARE offices that no information was available about USFHP.

Table 5.4

Selected Focus Group Comments Regarding Enrollment in USFHP

Site	Comment
Bayley Seton	They have never heard of the plan.
	Materials should be sent to people when they retire.
	This is not a very major program; people may never find out about it; it is not even advertised in military service newsletters.
	The program is not practical for many people as it only covers South New Jersey.
	The program has a 30-day open enrollment.
Brighton Marine	They never heard about it.
	They need more information.
	Many have insurance through employers.
	It is hard to compare benefits across different MHS plans.
	Brighton Marine is inconvenient, though care is impressive.
	The military should do a better job of advertising worthwhile programs; people do not want to take time to compare.
PacMed	Non-enrollees don't have information, so they don't enroll.
	Co-payments and premiums may be a deterrent.
	Employer-sponsored insurance is a big competitor, but at retirement USFHP may be more attractive.
	Information is a problem. Mailing lists may have to be updated.
	Coverage area is an important factor.

Impact of Continuous Open Enrollment

Following the discussions summarized above, participants were reminded of the open enrollment demonstration program and the role of the focus group in evaluating the demonstration. Few participants indicated that they had been personally affected by the previous practice of annual open enrollment, and more stated that their recruitment to the focus group was the first they had learned of the demonstration. One respondent indicated that annual open enrollment caused a delay in obtaining needed care. Several respondents reported going without health insurance altogether as a result of annual open enrollment. The majority of participants believed that continuous open enrollment was beneficial for military retirees. Few had no opinion, and several also indicated that open enrollment may alleviate difficulties noted above in marketing, because under annual open enrollment, information about USFHP seemed to be provided only once a year. (See Table 5.5).

Many participants simply did not understand why annual open enrollment might be preferred over continuous open enrollment. This perception was particularly acute among the non-Medicare-eligible population, because these participants are eligible for TRICARE, which has always practiced continuous

Table 5.5

Selected Focus Group Comments Regarding Continuous Open Enrollment

Site	Comment
Bayley Seton	It is annoying to have to wait.
	Open enrollment could affect access to care.
	Participant needed to be on COBRA until open enrollment and delayed knee surgery.
	It was inconvenient to wait five months to join USFHP.
	Concerns about premiums going up if open enrollment is adopted.
Brighton Marine	Participant wanted to join right away, but went without coverage for months.
	Open enrollment gives people the opportunity to get coverage when they need it.
	Participant lost coverage for a specific treatment and immediately joined USFHP
	A limited annual open enrollment does not reduce procrastination.
PacMed	Recruitment letter was the first indication of continuous open enrollment.
	Participant went without care for several months under annual open enrollment (2 respondents).
	Continuous open enrollment is a good idea.

open enrollment. Several Medicare-eligible participants indicated that more people would likely enroll in USFHP during continuous open enrollment because of recent trends in escalating Medigap costs. When asked whether they felt that continuous open enrollment might cause some individuals to procrastinate because they are not given a deadline, most participants responded that the cost of USFHP is low enough to encourage enrollment. Several mentioned that having coverage was important because of the need for "peace of mind," especially for the Medicare-eligible. Participants were specifically asked whether continuous open enrollment would cause people to postpone enrolling until they were sick and needed care. Across all groups, the overwhelming response was that people would not do this. Instead, they would enroll when they had the opportunity. Again, stability and certainty were given as reasons for not waiting.

Enrollment—Analysis of Enrollment Data

Conceptual Overview

Increased enrollment in USFHP resulting from the demonstration could arise in three possible ways.[1] First, beneficiaries under age 65 could shift from TRICARE to

[1]In general, only the time between enrollment under the demonstration and the date when the next one-month open enrollment period would have occurred (six months on average) can be attributed to the demonstration, since beneficiaries would have been able to enroll in USFHP at that time in any case. This may not always be true, however. In particular, we can imagine beneficiaries

USFHP. However, this is unlikely to have a major effect on costs, because USFHP capitation payments are required not to exceed what DoD would pay to cover comparable beneficiaries under TRICARE. Furthermore, as we have discussed, beneficiaries effectively can make this shift on a continuous basis under current enrollment rules; the continuous open enrollment policy would simply reduce the administrative costs of shifting.

Second, beneficiaries under age 65 who would otherwise not enroll in any DoD program might enroll in USFHP because of the demonstration. This would raise the costs to DoD and the federal government by the full amount of the capitation payment for that beneficiary.

Third, Medicare-eligible beneficiaries who would otherwise participate in Medicare might enroll in USFHP because of the demonstration. This would decrease costs to HCFA and increase DoD costs: as discussed above, USFHP enrollees may not concurrently use their Medicare benefits, but DoD is not reimbursed by HCFA for USFHP capitation payment.[2] In addition, USFHP capitation payments are required not to exceed what HCFA would pay to cover comparable beneficiaries under Medicare, plus an additional amount to cover the USFHP benefits that are not covered by Medicare (e.g., preventive services and prescription drugs). Increased USFHP enrollment by Medicare-eligible beneficiaries is thus also likely to increase net costs to the federal government.[3] Because this report was due before a full year had elapsed since the onset of the demonstration at the three demonstration sites, we had access only to enrollment for a partial year under the demonstration. Furthermore, for reasons described above, we were not formally able to ascribe the enrollment patterns we observed to the demonstration per se.

Empirical Results

Table 5.6 summarizes total enrollment in USFHP in FY1996 and in September 2000.[4] For comparability with the 1996 enrollment data, Table 5.6 includes all

---

who experience unexpected negative health shocks. Under the annual enrollment regime, they could not join USFHP until the next enrollment period. They would seek care under their current insurance, thus establishing relationships with providers that they may be reluctant to abandon to enroll in USFHP during the next enrollment period. In this case, if they cannot enroll in USFHP immediately via continuous open enrollment, they may not enroll at all for the foreseeable future. We discuss these issues in greater detail below.

[2]General Accounting Office (1996). Medicare Costs and Other Issues May Affect Uniformed Services Treatment Facilities' Future. Report GAO/HEHS-96-124.

[3]This would be true only if the extra USFHP benefits led to lower total health care costs, which is doubtful given the literature on the outcomes of preventive service use.

[4]Source for FY1996 data: General Accounting Office (1996). Medicare Costs and Other Issues May Affect Uniformed Services Treatment Facilities' Future. Report GAO/HEHS-96-124. Source for FY2000 data: authors' calculations based on data provided by the Iowa Foundation for Medical Care.

Table 5.6

USFHP Enrollment, 1996 and 2000

Site	1996	2000	% Change
Johns Hopkins	23,881	21,050	–11.9%
Brighton Marine	11,892	8,958	–24.7%
CHRISTUS Health	27,054	17,537	–35.2%
Sisters of Charity	15,772	12,940	–18.0%
PacMed	20,048	12,833	–36.0%
Martin's Point	18,795	16,948	–9.8%
Fairview	6,570	3,904	–40.6%
Total	124,012	94,170	–24.1%

USFHP enrollees in the figures for 2000, including dependents of active-duty personnel. Overall, the data indicate that enrollment at each of the demonstration sites has declined substantially in the past several years. This is consistent with information provided by USFHP staff, who indicated that overall enrollment in USFHP has stabilized or fallen in recent years, and has never reached the maximum growth rate of 10 percent per year currently permitted by law.

Table 5.7 provides more detail on the three demonstration sites as of September 2000.[5] All data in this table exclude dependents of active-duty personnel, since they were not covered by the demonstration. The first panel lists the eligible population, including a breakdown by whether beneficiaries have reached age 65. The second panel lists total enrollment, along with the membership rate (enrollees divided by eligibles). The third panel lists new enrollment during the demonstration period, along with the fraction of all enrollees who enrolled under the demonstration (new enrollees divided by total enrollees) and the uptake rate under the demonstration (new enrollees divided by non-enrolled eligibles).

The data in Table 5.7 suggest that beneficiaries age 65 and over are underrepresented among new enrollees compared with the overall enrolled population. For instance, across the three demonstration sites, 36.5 percent of beneficiaries enrolling during the demonstration period were over 65, compared with 49.8 percent of all enrollees.

Analysis of monthly enrollment patterns since the onset of the demonstration indicated that enrollment was approximately uniformly distributed by month at Brighton Marine and Bayley Seton; at PacMed, enrollment bunched in April 2000, coincident with the beginning of the benefit year under the previous enrollment policy.

[5] Source: authors' calculations based on data provided by the Iowa Foundation for Medical Care.

Table 5.7

Enrollment Under the Continuous Open Enrollment Demonstration

Since	Brighton Marine April 1, 2000		Bayley Seton January 1, 2000		PacMed March 1, 2000		Total	
	N	%	N	%	N	%	N	%
Eligibles	68,327		76,912		86,372		231,611	
Under age 65	30,025	43.9	32,522	42.3	25,751	29.8	88,298	38.1
Age 65 and over	38,312	56.1	44,402	57.7	60,623	70.2	143,337	61.9
Enrollees	7,792		9,588		12,240		29,620	
Under age 65	3,973	51.0	5,129	53.5	5,777	47.2	14,879	50.2
Age 65 and over	3,819	49.0	4,459	46.5	6,463	52.8	14,741	49.8
Enrollees/Eligibles		11.4		12.5		14.2		12.8
Under age 65		13.2		15.8		22.4		16.9
Age 65 and over		10.0		10.0		10.7%		10.3
Enrolled during demonstration	307		651		459		1,417	
Under age 65	192	62.5	415	63.7	293	63.8	900	63.5
Age 65 and over	115	37.5	236	36.3	166	36.2	517	36.5
New/Enrollees		3.9		6.8		3.8		4.8
Under age 65		4.8		8.1		5.1		6.0
Age 65 and over		3.0		5.3		2.6		3.5
New/Non-enrolled eligibles		0.5		1.0		0.6		0.7
Under age 65		0.7		1.5		1.5		1.2
Age 65 and over		0.3		0.6		0.3		0.4

For reference, Table 5.8 gives information on the four non-demonstration sites as of September 2000.[6] All data in this table exclude dependents of active-duty personnel, since they were not part of the target population for this evaluation. The first panel lists total enrollment. The second panel lists new enrollment at and after the most recent (FY2000) annual enrollment period, along with the fraction of all enrollees who enrolled under the demonstration (new enrollees divided by total enrollees). We did not obtain data on the eligible population in the four non-demonstration USFHP sites.

The data in Table 5.8 suggest that the rate of new enrollment is slightly higher, on average, in the non-demonstration sites than in the demonstration sites (6.4 percent versus 4.8 percent, respectively), despite the longer period over which enrollment could accumulate in the demonstration sites. Although these comparisons should be interpreted with caution because of the many differences between the various USFHP sites, these data do suggest that the demonstration has had minor or even no effects on enrollment to date. We note that, while a lower fraction of new enrollees in the non-demonstration sites are age 65 and over (21.1 percent, compared with 36.5 percent at the demonstration sites), the fraction of all enrollees who are 65 and over is also lower at the non-demonstration sites (33.1 percent, compared with 49.8 percent at the demonstration sites).

As we have discussed, we have no definitive way to identify what part, if any, of the new enrollment since the onset of the demonstration is attributable to the continuous open enrollment policy per se. In addition to the reasons we discussed in the first section, the implementation of the network expansions at Bayley Seton and Brighton Marine—which were explicitly intended to attract new enrollees—overlapped with the demonstration period. Indeed, analysis of the geographic distribution of enrollment suggested that half or more of the new enrollees at those two sites resided in areas where providers had recently been added to the respective USFHP programs.

Effects of the FY2001 Defense Authorization Act

As discussed above, the new insurance benefits that were included in the FY2001 Defense Authorization Act will substantially alter the relative attractiveness of the USFHP program for Medicare-eligible beneficiaries. Although many of the specific details are currently unclear, the act certainly provides prescription drug

---

[6] Source: authors' calculations based on data provided by the Iowa Foundation for Medical Care.

Table 5.8

Enrollment Under the Continuous Open Enrollment Demonstration

Since	Hopkins June 1, 2000		CHRISTUS April 1, 2000		Martin;s Pt. June 1, 2000		Fairview May 1, 2000		Total	
	N	%	N	%	N	%	N	%	N	%
Enrollees	17,842		16,309		13,596		3,112		50,859	
Under age 65	12,576	70.5%	9,793	60.0%	9,297	68.4%	2,340	75.2%	34,006	66.9%
Age 65 and over	5,266	29.5%	6,516	40.0%	4,299	31.6%	772	24.8%	16,853	33.1%
Newly Enrolled*	1,267		654		1,106		235		3,262	
Under age 65	991	78.2%	566	86.5%	810	73.2%	208	88.5%	2,575	78.9%
Age 65 and over	276	21.8%	88	13.5%	296	26.8%	27	11.5%	687	21.1%
New/Enrollees		7.1%		4.0%		8.1%		7.6%		6.4%
Under age 65		7.9%		5.8%		8.7%		8.9%		7.6%
Age 65 and over		5.2%		1.4%		6.9%		3.5%		4.1%

*At and after most recent annual enrollment period.

coverage that is comparable to the USFHP benefit. Medicare-eligible beneficiaries will thus be able to combine prescription drug benefits with fee-for-service Medicare and avoid the managed-care restrictions imposed by USFHP. Furthermore, it seems likely that TRICARE will serve as a wrap-around Medigap policy, covering most or all out-of-pocket costs under fee-for-service Medicare. In sum, it seems plausible that the new benefits will lead to a net decrease in USFHP enrollment by Medicare-eligible beneficiaries.

Adverse Selection

To our knowledge, neither DoD nor USFHP has expressed concern that continuous open enrollment would lead to adverse selection into USFHP by beneficiaries under age 65, mainly because such beneficiaries already have the opportunity to enroll in TRICARE Prime at any time. Furthermore, as discussed above, the policy of "transferability" represents a de facto continuous open enrollment policy, independent of the demonstration—although at some administrative cost to beneficiaries, DoD, and USFHP, since beneficiaries need to enroll in TRICARE Prime with a Managed Care Support Contractor before transferring to USFHP.

Therefore, we focused on the potential for adverse selection among Medicare-eligible beneficiaries. As mentioned earlier, adequate data to evaluate this issue empirically were not available by the deadline for this report.[7] Such data should become available over the coming months, should DoD wish to revisit this issue.

Risk to DoD

Given the history and design of the USFHP program and the nature of the demonstration, the most plausible source for adverse selection resulting from the continuous open enrollment policy is beneficiary enrollment immediately at or after an acute health shock requiring intensive follow-up care (e.g., a stroke or heart attack). In general, such enrollment is precluded under the annual open enrollment policy, since unenrolled beneficiaries would have to wait until the next designed enrollment period.

However, as we have discussed, it is unclear whether the continuous open enrollment policy leads beneficiaries to enroll in USFHP earlier or later than

[7]For reference, 517 beneficiaries age 65 and over had enrolled by September 2000, with nearly half of these enrolling at Bayley Seton (where TMA is concerned about data quality). Of these, 327 had at least one medical or pharmacy claim during the first three months of enrollment, and 19 had at least one inpatient claim.

otherwise. Earlier enrollment under the conditions just described would certainly increase DoD's costs by increasing the period over which DoD is paying for USFHP participation. Furthermore, beneficiaries would enter the program in relatively poor health, which in principle might lead to higher capitation payments from DoD under risk adjustment.

In practice, however, DoD is unlikely to bear most of the cost of such adverse selection, for several reasons. First, under the ACG risk-adjustment methodology, differences in the distribution of diagnosed morbidity between the USFHP population and the general Medicare population are used to adjust DoD capitation payments in subsequent years. These adjustments, which are made without regard to health service use per se, are unlikely to offset the costs incurred by USFHP in the initial enrollment period (nor are they designed to do so), and there is currently no contractual provision for reimbursing USFHP retroactively to offset unexpectedly high health service use. Finally, at some sites, the current USFHP contracts limit DoD's exposure to higher capitation rates resulting from adverse selection.

In contrast, later enrollment—e.g., because beneficiaries forgo enrollment until they need medical care, which USFHP personnel at each demonstration site speculated was happening—has ambiguous effects on DoD's costs. The shorter period of enrollment lowers DoD's costs. In principle, this effect may be at least partially offset by a higher capitation payment triggered by the poor health of the beneficiary at the time of enrollment. For the reasons we have just discussed, however, we think that such offset is unlikely under the current USFHP contracts.

As noted above, neither site visit nor focus group participants identified any specific characteristics of the USFHP program that would make it particularly attractive (relative to other plans) in the event of an acute episode of health care need, assuming people were already enrolled in another health insurance plan.[8,9] The attraction of USFHP that appears most likely to motivate adverse selection is the pharmacy benefit—but, as we have discussed, the continuous open enrollment policy is unlikely to have a major effect on that source of adverse selection.

---

[8]This finding is specific to the demonstration sites and might differ at other USFHP sites.

[9]Furthermore, evidence from the nonmilitary Medicare population suggests that beneficiaries actually leave Medicare HMOs for fee-for-service Medicare when they have a relatively acute demand for intensive medical care, i.e., the kind of utilization patterns that are particularly expensive.

USFHP's Risk

Perhaps surprisingly, given USFHP's request for the demonstration, it seems likely that the financial risk to USFHP of adverse selection resulting from the continuous open enrollment policy is substantially higher than the risk faced by DoD. In particular, for the reasons described above, we find it unlikely that the risk-adjustment mechanisms used by DoD and USFHP to set capitation rates would fully offset the costs of an acute episode of health care use. USFHP would face a financial risk from adverse selection whether the demonstration induced beneficiaries to enroll earlier or later than otherwise.

We note that additional net costs resulting from adverse selection could, in principle, threaten the long-term financial viability of USFHP. In this case, the mechanisms for setting capitation rates might be the subject of future negotiations.

Effects of the FY2001 Defense Authorization Act

The new benefit structure may significantly alter the incentives for adverse selection into USFHP. In addition to the effects of the new pharmacy benefit, discussed above, TRICARE's function as a Medigap policy may make fee-for-service Medicare (rather than USFHP, if that were indeed currently the case) particularly attractive in the event of an acute episode of health care need.

Finally, if Medicare-eligible military beneficiaries become eligible for TRICARE Prime—which is currently an option only under the TRICARE Senior Prime demonstration program—it seems plausible that the policy of "transferability" will extend to Medicare-eligible beneficiaries. This would create the same de facto policy of continuous open enrollment into USFHP for Medicare-eligible beneficiaries as currently exists for beneficiaries under 65, with the only barrier being the administrative hassles of transferring between programs.

6. Conclusions

Consequences and Costs for Beneficiaries

Our focus group findings and our analysis of the broader insurance market in which USFHP operates suggest that the continuous open enrollment policy has potential benefits for beneficiaries. For instance, the policy increases access to the program for eligible beneficiaries under age 65 but eliminates the need to enroll in TRICARE Prime first and then transfer to USFHP. More important, continuous open enrollment decreases the chance that eligibles age 65 and over will go without supplemental insurance or purchase relatively expensive Medigap plans. The potential value of this policy is likely to increase as Medicare HMO plans close; since the benefit years of such plans do not in general align with the USFHP benefit year, continuous open enrollment permits military beneficiaries who lose their Medicare HMO coverage to enroll in USFHP with no delay. This benefit of continuous open enrollment is likely to decline in relative terms with the introduction of the new benefits under the FY2001 Defense Authorization Act.

Our investigation revealed no drawback to continuous open enrollment for beneficiaries, unless the policy impaired the financial viability of the USFHP programs.

Consequences and Costs for DoD

The most likely source of increased costs to DoD is increased enrollment rather than adverse selection. The primary reason for this observation is our assessment that the current system for setting capitation rates places most risk from adverse selection resulting from continuous enrollment on USFHP.

Based on the focus group results, it seems plausible that eligible beneficiaries not currently enrolled in USFHP are not enrolled because they either do not know about the plan or are concerned about access and USFHP provider turnover. Site visits revealed challenges in marketing the program in the 30-day open enrollment window, and focus groups confirmed the lack of awareness of the program among the eligible population. Continuous open enrollment will permit more-comprehensive marketing efforts, which will likely increase knowledge. At the same time, network expansions will address another barrier to enrollment identified by focus group participants—namely, concern about the affiliated providers and facilities. Design and data limitations and contemporaneous policy changes precluded our ability to specifically identify whether the continuous open enrollment policy had an effect on overall USFHP enrollment. Nevertheless, it seems plausible that a continuous open enrollment policy will somewhat increase enrollment relative to the current policy. At the same time, given the new benefits under the FY2001 Defense Authorization Act, we expect that most new USFHP enrollees would otherwise be enrolled in a different federal health insurance program, e.g., TRICARE (for those under age 65, as is currently the case), or Medicare plus TRICARE plus the National Mail Order Pharmacy. Thus, the net costs to DoD of increased USFHP enrollment will depend on the details of how these new benefits are structured, but the net costs to the federal government as a whole are likely to be small.

Consequences and Costs for USFHP

The net consequences for USFHP are ambiguous and largely outside the scope of this evaluation. In requesting a policy of continuous open enrollment, the USFHP program directors apparently expected that savings from more efficient marketing and increased revenue from higher enrollment would more than offset potential losses resulting from adverse selection. If this is incorrect, the continuous enrollment policy and other aspects of the contracts between DoD and USFHP may be the subject of future negotiations. In any case, the FY2001 Defense Authorization Act, enacted since the onset of the demonstration, has substantially changed the policy environment in which USFHP operates, and the challenges presented are likely to have a more significant long-term effect on the USFHP program than would continuous open enrollment.

Limitations

Many of this evaluation's limitations have been discussed above. The design of the demonstration limited our ability to identify its effects definitively, because of the lack of an obvious control population and the many contemporaneous changes in various USFHP programs and in the broader policy environment in which the demonstration was being conducted. The time between the onset of the demonstration period and the deadline for this report was also relatively short, which particularly limited our ability to analyze health care use under the demonstration. (Such analyses would be possible once sufficient numbers of beneficiaries—particularly Medicare-eligible beneficiaries—have enrolled under the continuous open enrollment policy.) Finally, we had limited ability to account for possible differences between the three demonstration sites and the four other USFHP sites, respectively. However, we believe that our conclusions about the consequences and costs of the continuous open enrollment policy should be generally applicable to each USFHP site.

Recommendations

We recommend that continuous open enrollment be extended to all USFHP sites on a permanent basis, conditional on this policy being acceptable to the USFHP programs themselves. Short of this result, we recommend that the enrollment policies of USFHP match those of TRICARE Prime, for beneficiaries who are eligible for both programs.

Appendix

A. Intended Analyses of Health Care Use and Costs

We obtained and intended to analyze data on USFHP medical claims and encounters from the Iowa Foundation to help evaluate the possible costs to DoD of the continuous open enrollment policy. However, as we describe below, we concluded that the data currently available would not support valid and generalizable analyses. Thus we did not actually conduct the claims analyses described here.

The main concern expressed by DoD regarding the open enrollment policy was that it would promote adverse selection, particularly by Medicare-eligible beneficiaries. DoD focused specifically on this population because, historically, these beneficiaries have not been eligible for any other DoD health insurance.[1]

In general, adverse selection could take several forms. First, beneficiaries with relatively poor health status (e.g., those with chronic diseases) could disproportionately enroll in USFHP. This sort of selection could be driven by USFHP's relatively generous benefit package, especially regarding prescription drugs, and this is presumably why USFHP initially requested that capitation rates consider beneficiaries' health status. Such selection could occur with or without the demonstration; indeed, since the prevalence of chronic disease exceeds the incidence, it seems plausible that the bulk of such selection would be independent of the demonstration.

Second, beneficiaries experiencing an unanticipated negative heath shock (i.e., a myocardial infarction or some other unanticipated acute event) could subsequently enroll in USFHP under the demonstration, at a time when they would not have been eligible to enroll. Such selection could take several forms, with ambiguous implications for DoD's costs. For instance, beneficiaries who would otherwise have enrolled in USFHP during a given one-month open enrollment period might delay enrollment until later in the year, when a particular health need arose. In this case, DoD would be responsible for

[1]Source: http://www.tricare.osd.mil/tricare/news/faqs.htm.

capitation payments over a shorter period of time under the demonstration. In contrast, the capitation rate could be higher since (under this scenario) the beneficiaries entered USFHP in worse health under the demonstration; in practice this would also depend on the specific risk-adjustment methods used by DoD and USFHP to determine capitation rates.

Alternatively, beneficiaries who would otherwise have waited to enroll in USFHP in the subsequent one-month open enrollment period would enroll earlier, when a particular health need arose. In this case, DoD would be responsible for capitation payments over a longer period of time; again, the capitation rate would depend on the specific risk-adjustment methods.[2]

Thus, as part of the evaluation, we wanted to look for evidence that enrollment decisions under the demonstration were more strongly influenced by acute health care problems than were decisions made under the one-month open enrollment policy. We proposed to gather this evidence by comparing health care use in the first *n* months of enrollment for USFHP enrollees in the demonstration sites who had enrolled since the onset of the demonstration with that of individuals who had enrolled during the prior two one-month open enrollment windows. We were thus interested in estimating multivariate models

$$Y_t = g(X'_o\beta + S\theta + T'\delta + \varepsilon),\ \varepsilon \sim F(\varepsilon), \qquad t>0$$

where Y_t denotes the value of the outcome measure at follow-up period t, such as total health care costs over the period, or the probability that beneficiaries were hospitalized during the period; X_o' denotes a vector of individual characteristics that may be associated with Y_t, such as age, gender, and sponsor status (sponsor versus spouse); S' denotes a vector of dummy variables that represent site-specific fixed effects; T' denotes whether the beneficiary enrolled since the onset of the demonstration or previously; β, θ, and δ denote parameters to be estimated (δ is the intervention effect); ε denotes the error term; $F(\varepsilon)$ denotes the error distribution; and $g[\cdot]$ denotes the link function for the model. In this application, the models take linear or logistic form depending on the choice of g and F.

To help examine these issues, we obtained data on USFHP medical and pharmacy claims since 1997 from the Iowa Foundation for Medical Care. However, a number of factors led us to conclude that the available data did not

[2]Furthermore, under the scenario described in footnote 1 of Section 5, it might be appropriate to consider the costs of USFHP participation in subsequent years also to be attributable to the continuous open enrollment policy, e.g., because beneficiaries would not have enrolled in USFHP in the foreseeable future if they had not been able to enroll around the time of the health shock.

permit appropriate quantitative analyses, given the relatively low frequency of outcomes, such as inpatient hospitalization in the USFHP population (even among Medicare-eligible beneficiaries), and the highly skewed distribution of medical costs. Most important, as we describe in greater detail below, the available sample size was limited by a combination of the continuous nature of enrollment under the demonstration, relatively low enrollment under the demonstration period, a short time period between the onset of the demonstration and the deadline for this report, and a lag of at least three months before claims data could be considered complete (see Table 5.7). In addition, TMA and the Iowa Foundation indicated to us that data quality from one of the three demonstration sites was relatively poor, with further limited sample size.

B. Organization and Delivery of Care at USFHP Demonstration Sites

Although the provision of care by USFHP originally centered on former public health hospital facilities, the organization and delivery of care has evolved differently in each of the sites in ways that influence overall enrollment in the program and the potential for adverse selection that results from continuous open enrollment.

Brighton Marine

In 1981, a nonprofit association of retired military personnel took ownership of the Brighton Marine hospital and currently functions as a realty holding company. Members of the association serve on the Brighton Marine governing board, which in turn contracts with St. Elizabeth's, a Catholic hospital located across the street from the Brighton Marine building, to provide outpatient services on the premises. The Brighton Marine Board provides leadership, strategic planning, and moral guidance for St. Elizabeth's, with particular emphasis on issues related to care for Medicare-eligible military beneficiaries.

The Brighton Marine building has been extensively renovated and now serves as the flagship ambulatory care clinic. St. Elizabeth's negotiates and administers the USFHP contract and directly staffs and manages bricks-and-mortar facilities of St. Elizabeth Hospital, the Brighton Marine ambulatory care clinic, and a remote clinic on Hanscom Air Force Base (with a primary care clinic and five specialty clinics) affiliated with USFHP. All other care rendered to USFHP enrollees is provided by network physicians who practice in private offices. St. Elizabeth's also participates in other managed-care provider networks, including one sponsored by Tufts University. However, until recently, the Brighton Marine building and St. Elizabeth's Hospital have been closely identified with the military, which site visit participants indicated has limited their market share in the civilian community.

Traditionally, the Brighton Marine Clinic and St. Elizabeth's Hospital have provided the bulk of care to USFHP enrollees, using a tightly managed primary care provider/gatekeeper model. USFHP enrollees account for 60 to 70 percent of outpatient visits at the Brighton Marine building and 15 percent of inpatient stays (formerly as high as 80 percent) at St. Elizabeth's.

More recently, the need to maintain enrollment has prompted the USFHP program to create a broader network of affiliated providers. The program is now affiliated with the University of Massachusetts Hospital, Children's Hospital of Boston, a health system located on Cape Cod, and a provider network that serves the Newport, Rhode Island, area. Despite efforts to have enrollees choose primary care providers near their home, many enrollees are emotionally connected to the physical facilities of the Brighton Marine Clinic and St. Elizabeth's Hospital, and some travel long distances to use them. This use pattern has been costly for Brighton Marine, because when these members require emergency care, they end up using non-network facilities close to home.

A map of the Brighton Marine USFHP service area is included in Appendix J. A second map distinguishes between the historical Brighton Marine service area and areas in which the Brighton Marine USFHP has added providers in the last several years.

Bayley Seton

The USFHP sponsored by Bayley Seton is divided into three distinct service areas, defined by geographic barriers. The first is located on Staten Island and consists of an ambulatory care clinic located in the former Bayley Seton Hospital, a group-based health center called Bayridge, and inpatient services and specialty care clinics at St. Vincent's Hospital located several miles from the clinics at Bayley Seton. These three facilities serve enrollees who reside on Staten Island and in Brooklyn. Similar to Brighton Marine in Boston, the Bayley Seton USFHP follows a primary care physician-based gatekeeper model. As at Brighton Marine in Boston, many current enrollees have used the Bayley Seton clinics and former public health hospital for more than 25 years and are reluctant to use other providers and facilities that lack a similar connection to the military. The Bayley Seton USFHP also runs a primary care–based health center in Nassau County on Long Island.

Efforts are currently under way to expand the number of network providers on Suffolk County. The program in New Jersey is based on a preferred provider or PPO model that uses a physician network organized by a third party administrator (Qualcare). Although Bayley Seton historically has included some enrollees living in New Jersey, most of the providers in New Jersey represent recent expansions of Bayley Seton's provider network that have occurred in the last few years. Unlike the enrollees on Staten Island and Brooklyn, the majority of enrollees in New Jersey—particularly in the central and southern parts of the state—are new to the USFHP network.

Currently, Bayley-Seton has no other managed-care contracts but plans to compete for contracts in the future. St. Vincent's Hospital is affiliated with organizations that have Medicare risk contracts. However, most non-USFHP business conducted at the three facilities is fee-for-service. Recently, St. Vincent's Hospital merged with two other Catholic health centers: Catholic Health Center, which serves Brooklyn and Queens, and St. Vincent's Catholic Medical Centers of New York. These new affiliations should make the network more accessible to eligibles who live in the other four boroughs of New York City.

A map of the Bayley Seton USFHP service area is included in Appendix J.

PacMed

PacMed has approximately 120 salaried providers who see only PacMed patients (approximately 80 primary care providers and 40 specialists, working in 12 clinical sites), as well as staff physical therapists and mental health social workers. In addition, PacMed contracts with approximately 500 community providers, primarily medical subspecialists. Most of PacMed's clinics are rented. PacMed staff and network clinicians provide approximately 250,000 patient visits per year.

PacMed owned its own inpatient facility until 1987, when it was sold and ultimately closed (the building is now used primarily as the corporate headquarters for the Internet retailer Amazon, although PacMed still rents clinic space and administrative space in the basement). PacMed contracts with Swedish Hospital, the largest inpatient facility in Seattle, to provide inpatient care to its members (PacMed employs hospitalists at Swedish Hospital). PacMed operates pharmacies in four of its clinics, and it also contracts with a local chain of retail pharmacies.

In addition to operating the USFHP program, PacMed has full-risk managed-care contracts with several Medicare HMOs. At the time of the site visit, about 60 percent of PacMed's total business came through managed-care contracts. USFHP is the largest single piece, but PacMed has a total managed-care enrollment of approximately 45,000 covered people. The remaining 40 percent of PacMed's business is commercial fee-for-service and charity care (PacMed provides between $1 million and $2.5 million in charity care per year). PacMed anticipated that its service and contract mix would change in 2001, since several of the Medicare HMOs were scheduled to leave the Seattle market.

According to PacMed staff, approximately 47 percent of PacMed's USFHP enrollees are age 65 or older, among the highest percentages in the USFHP program. In addition, they stated that PacMed has the highest fraction of Coast Guard enrollees and the smallest fraction of active-duty dependents among all USFHP locations.

A map of the PacMed USFHP service area is included in Appendix J.

C. Medicare Managed-Care Market in USFHP Demonstration Sites

Over the past decade, the proportion of Medicare beneficiaries enrolled in managed-care plans has grown substantially, from roughly 5 percent in 1993 to 17 percent in 2001. Although these plans generally restrict provider choice, beneficiaries have been attracted to Medicare managed-care plans because of low premiums and out-of-pocket costs and coverage of a range of services not covered by Medicare, including prescription drugs. The cost and generosity of Medicare managed-care plans vary widely across the country, depending on local market conditions and payment rates set by HCFA, which reflect the cost of providing fee-for-service Medicare benefits in particular counties. Over the past two years, reductions in payments to managed-care plans in areas of the country with traditionally generous coverage and rapid increases in the use and cost of prescription drugs have led to premium increases, benefit cuts, and Medicare HMO plan closures.

Table C.1 provides descriptive information about the Medicare managed-care markets in one county in each of the three demonstration sites. We chose these three counties as examples because site visit discussions suggested that they contain a high concentration of Medicare-eligible USFHP enrollees. Recent network expansions generally have taken place outside of these counties. However, these network expansions generally have occurred in less densely populated areas where Medicare managed-care markets are less developed and competition is less intense as a result of fewer plans, higher premiums, and less-generous benefits.

Overall, Table C.1 suggests that the three USFHP demonstration sites compete for enrollees in well-developed Medicare managed-care markets. Medicare managed-care enrollment is roughly equal to or above national averages in the USFHP counties. Compared with the median Medicare managed-care plan, USFHP charges higher enrollment fees (versus premiums) and higher office visit co-payments. At the same time, the USFHP drug benefit is substantially more generous than that offered by typical Medicare HMO plans.

However, the comparisons between USFHP and Medicare managed-care plans are highly variable at the local level. The Bayley Seton plan operates in the most competitive Medicare managed-care market, reflecting Medicare's substantially higher per-enrollee payment to managed-care plans operating in the area. Five of the seven plans operating in Richmond County (Staten Island), New York, have premiums of $10 or less per month, almost half that of USFHP and substantially less expensive than plans in the other two sites. Although less generous than the USFHP drug benefit, the unlimited coverage of generic drugs in Richmond County is substantially more generous than that offered by Medicare managed-care plans in the other two market areas. By contrast, the PacMed and Brighton Marine plans offer a substantially more generous drug benefit at a lower premium compared with the Medicare managed-care plans operating in their market areas. Recent and impending benefit reductions in Medicare managed-care plans suggest that the relative generosity of USFHP plans is likely to increase over time in each of the three markets.

Brighton Marine

Competition from the regional Managed Care Support Contractor (Sierra) has been slow to develop in the Boston area. Sierra has to date focused its efforts in the Washington, DC, area, where concentrations of military beneficiaries are substantially greater. The network that Sierra has created in the Boston area currently lacks high-profile hospitals and medical groups. Traditionally, Brighton Marine has had a highly collaborative relationship with the MTFs located at Hanscom and Newport based on the receipt of a large number of inpatient referrals prior to 1993 and the fact that both MTFs and Sierra offer very limited pediatric services. However, over the past year the relationship has become more competitive, both because space has been available at MTFs and because Sierra recognizes that Brighton Marine has enrolled a large share of its potential market.

Site visit participants did not foresee substantial increases in enrollment resulting from the termination of Medicare risk contracts in the area. They expressed the belief that patients choose Medicare managed-care plans because of their own doctors' affiliation with the plans. They speculated that managed-care enrollees would simply switch to fee-for-service Medicare and continue to see the same physician.

Site visit participants considered the potential for adverse selection resulting from continuous open enrollment to be limited. Participants speculated that although Brighton Marine and St. Elizabeth's Hospital (and the Tufts University network more generally) provide high-quality care, they are simply overshadowed in the market by plans affiliated with top-tier Boston hospitals, such as Massachusetts General. They perceived it as unlikely that individuals

Table C.1

Medicare Managed-Care Market and Plan Descriptions in USFHP Continuous Open Enrollment Demonstration Sites Compared with Typical National Plan and Rates

	Base Payment 2000[b]	Medicare Managed-Care Market Share 2000[c]	Plan Name	Monthly Premium 2001	Drug Benefit	Office Visit Co-payment
Median National Plan and Rates[a]	$479	17%		$0	83 % of offer a drug benefit and 47 % of plans with benefit have limit of $3,000 or less per year (1999)	$5
USFHP	NA	NA	All Sites	$19[d]	$9/30-day supply	$12
King County, WA	$482	24%	Group Health Cooperative of Puget Sound (High Option)	$194	Unlimited formulary drugs + $10/30 day supply	$5
			Group Health Cooperative of Puget Sound (Basic)	$54	None	$5
			Options Health Care, Inc.	$54	None	$5
			Pacificare of Washington, Inc.	$30	None	$10
Middlesex County, MA	$558	26%	Blue Cross and Blue Shield of Massachusetts	$85	$600/year + $8–$15/30-day supply	$5
			Fallon Community Health Plan (High Option)	$60	$175/"period" for formulary drugs	$10
			Fallon Community Health Plan (Basic)	$9	None	$10
			Harvard Pilgrim Health Care	$35	$600/year + $5–$25/30-day supply	$5
			Tufts Associated HMO, Inc.	$25	$600/year + $5–$25/30-day supply	$5

Table C.1 (continued)

	Base Payment 2000[b]	Medicare Managed-Care Market Share 2000[c]	Plan Name	Monthly Premium 2001	Drug Benefit	Office Visit Co-payment
Richmond County, NY	$814	33%	Aetna U.S. Health Care	$10	Unlimited generics + $500/year for other drugs + $10–$15/30 day supply	$10–$15
			Empire Blue Cross-Blue Shield	$0	$1,000/year + $5–$15/30-day supply	$10
			Health First Health Plans, Inc.	$25	Unlimited generics + $500/year for other drugs + $10–$15/30-day supply	$10
			HIP of Greater New York	$57	Unlimited generics + $500/year for other drugs+ $5–$15/30-day supply	Not shown
			Oxford Health Plans (New York) Inc.	$0	Unlimited generics + $7/30-day supply (nonformulary generics only)	$15
			Physicians Health Service of New York, Inc.	$0	Unlimited generics + $1000/year for other drugs + $7–$25/30-day supply	$10
			United Health Care of New York Inc.	$0	Unlimited generics + $7/30-day supply (non-formulary generics only)	$5

[a]Source: Langwell, K., C. Topoleski, D. Sherman, "Analysis of Benefits Offered by Medicare HMOs, 1999: Complexities and Implications," August 1999, Kaiser Family Foundation (Grant #97-1243).

[b]Payment rate information obtained from the Health Care Financing Administration web site: www.hcfa.gov/stats/hmorates/aapccpg.htm.

[c]Enrollment rate information obtained from the Health Care Financing Administration Web site: www.hcfa.gov/stats/ENROLL/DEFAULT.HTM.

[d]Monthly premium implied by $230 annual enrollment fee. Source: The Medicare Compare Database (www.medicare.gov/mphCompare/home.asp) as of January 7, 2001. When multiple plans were listed, the most expensive plan with drug benefits is shown. These plans may be closed to new enrollees.

would enroll in USFHP to receive care for limited and predictable acute events. They considered the biggest effect of continuous open enrollment to be on the timing of enrollment.

Bayley Seton

As in Boston, the Sierra network is weak in the New York area and has not focused on establishing member relationships within the USFHP provider network. Instead, Sierra has focused its marketing efforts on MTF cachement areas. Site visit participants suggested that Sierra has been getting stronger over time, but that in the near term, competition is more likely to come from the new health insurance benefits for DoD beneficiaries under the FY2001 Defense Authorization Act which could substantially affect the relative attractiveness of USFHP. At the same time, competition from Medicare HMOs was likely to fall over time as they raise premiums, reduce drug coverage, or close altogether (a particular issue in the New Jersey service area).

PacMed

PacMed staff mentioned that, historically, PacMed's covered population includes retired military beneficiaries (and their dependents) who joined via a Medicare HMO rather than by joining USFHP. They believed this was because the Medicare HMO coverage was part of an employer-sponsored retirement package. Some or all of these beneficiaries were expected to transfer to USFHP if the Medicare HMO left the Seattle market (although this expectation may have changed with the passage of the FY2001 Defense Authorization Act).

The PacMed service area overlaps with the cachement area for the Madigan MTF, which is approximately 50 miles from Seattle. Madigan is one of the sites currently participating the TRICARE Senior Prime demonstration, which may have attracted some beneficiaries away from USFHP. At the same time, the TRICARE Senior Prime program has led to a reduction in space-available care at participating MTFs for Medicare-eligible beneficiaries who do not enroll in the demonstration, and some of these beneficiaries may have been shifted towards USFHP.

D. USFHP Marketing

Brighton Marine

Site visit participants noted that most current enrollees were grandfathered into USFHP in 1993, and that the number of new members enrolled during the demonstration period would be small. Participants reported doing less marketing this year in the traditional service area. Instead of mailing large envelopes with descriptive information to all eligibles on the DEERS list, they mailed postcards that could be used to request further information. They felt that, given the poor quality of the address information contained in DEERS, this would reduce expenses. More-intense efforts were focused in marketing the program in the geographic areas covered by network expansions.

Bayley Seton

Site visit participants suggested that continuous open enrollment would allow better timing and geographic placement of marketing presentations and the use of a 35-day cycle for mailing marketing materials. Here, annual open enrollment occurred in the summer vacation season. They also noted that under continuous enrollment they could reduce expenses by using regular mail rather than priority mail. Participants speculated that continuous open enrollment would help Medicare-eligible enrollees save money because they could plan their enrollment to coincide with the termination of their supplemental (Medigap) plans and not have to pay premiums for coverage they were not using.

Site visit participants reported that marketing involved educating potential participants about the pros and cons of the program in town-hall settings and relied heavily on word of mouth from peers. They remarked that the closed nature of the service delivery model was not appropriate for all eligibles (e.g., snowbirds) and that continuous open enrollment would allow them to enhance the educational component of their marketing efforts in ways not possible before. They noted that their educational philosophy was reflected in their marketing materials, which include side-by-side comparisons with alternative coverage options.

Site visit participants estimated that 520 new members had enrolled as a result of continuous open enrollment (see Table 5.7), with the remainder of the new enrollment since the onset of the demonstration coming from new retirees, transfers from TRICARE Prime, and eligible beneficiaries moving into the area. Data were not available to verify this estimate independently, nor to identify these beneficiaries in the enrollment data from the Iowa Foundation.

PacMed

Comments by site visit participants regarding recruitment and enrollment activities were consistent with those described for the other two demonstration sites. Through September 2000, PacMed estimated that 261 new members had enrolled as a result of continuous open enrollment (see Table 5.7), with the remainder of the new enrollment since the onset of the demonstration coming from new retirees, transfers from TRICARE Prime, and eligible beneficiaries moving into the area. Data were not available to verify this estimate independently.

E. Sample USFHP Focus Group Recruitment Letter, Enrollees

Dear <<USFHP participant>>:

You are invited to participate in a focus group discussion of the Uniformed Services Family Health Plan (USFHP), your Military Health System benefits, and your health insurance needs.

As you may know, the Department of Defense is currently conducting the Uniformed Services Family Health Plan (USFHP) Continuous Open Enrollment Demonstration program in your area. This program permits retired Military Health System beneficiaries and their dependents to enroll in USFPH at any time, instead of during an annual one-month open enrollment period. This focus group is part of a study, sponsored by the Department of Defense, aimed at helping the federal government better understand the implementation of this program.

The Department of Defense has asked RAND to conduct this study. RAND is an independent, non-profit research organization with a national reputation for quality health care research.

The focus group will include USFHP participants and RAND researchers. During this meeting, you will have the opportunity to share your experiences with and thoughts about USFHP's open enrollment policy, your Military Health System benefits, and your health insurance needs, in a casual environment and with complete confidentiality. ***Your experiences and opinions are extremely valuable in helping the Department of Defense and Congress improve health insurance benefits for military retirees and their families.***

The focus group will be held on DATE from TIME1 until TIME2 at PLACE. <Continental breakfast OR snacks> will be provided. To compensate you for your time, all attendees will receive <$40 OR $50>. You were randomly selected from a list of USFHP participants. Although we hope you will join us, participation is voluntary. If you choose not to attend, it will not affect the benefits or medical care that you and your family receive. Please be assured that RAND will keep anything you say during the focus group strictly confidential, and that RAND will not release any information that can be linked to you.

A member of the research team will be contacting you by telephone to give you more details about this important event and answer any questions you may have about the study. You are also welcome to call us toll free at <1-800-XXX-XXXX>.

We hope that you will be able to join us for this important discussion.

Sincerely,

Michael Schoenbaum, PhD   C. Ross Anthony, PhD

F. Sample USFHP Focus Group Recruitment Letter, Non-Enrollees

Dear <USFHP eligible>:

You are invited to participate in a focus group discussion of the Uniformed Services Family Health Plan (USFHP), your Military Health System benefits, and your health insurance needs.

As you may know, the Department of Defense is currently conducting the Uniformed Services Family Health Plan (USFHP) Continuous Open Enrollment Demonstration program in your area. This program permits retired Military Health System beneficiaries and their dependents to enroll in USFHP at any time, instead of during an annual one-month open enrollment period. This focus group is part of a study, sponsored by the Department of Defense, aimed at helping the federal government better understand the implementation of this program.

The Department of Defense has asked RAND to conduct this study. RAND is an independent, non-profit research organization with a national reputation for quality health care research.

The focus group will include Military Health System beneficiaries eligible to participate in USFHP and RAND researchers. During this meeting, you will have the opportunity to share your thoughts about USFHP's open enrollment policy, your Military Health System benefits, and your health insurance needs, in a casual environment and with complete confidentiality. ***Your experiences and opinions are extremely valuable in helping the Department of Defense and Congress improve health insurance benefits for military retirees and their families.***

The focus group will be held on DATE from TIME1 until TIME2 at PLACE. <Continental breakfast or Snacks> will be provided. To compensate you for your time, all attendees will receive <$40 OR $50>.

You were randomly selected from a list of individuals who are eligible for USFHP. Although we hope you will join us, participation is voluntary. If you choose not to attend, it will not affect the benefits that you and your family receive, including your eligibility for the USFHP program. Please be assured that RAND will keep anything you say during the focus group strictly confidential, and that RAND will not release any information that can be linked to you.

A member of the research team will be contacting you by phone to give you more details about this important event and answer any questions you may have about the study. You are also welcome to call us toll free at 1-800-XXX-XXXX.

We hope that you will be able to join us for this important discussion.

Sincerely,

Michael Schoenbaum, PhD C. Ross Anthony, PhD

G. Sample USFHP Focus Group Confirmation Letter, Enrollees

DATE, 2000

Dear <Mr./Ms.> LASTNAME:

Thank you for agreeing to participate in a focus group discussion of the Uniformed Services Family Health Plan (USFHP), your Military Health System benefits, and your health insurance needs. RAND, an independent non-profit research organization, is conducting this focus group on behalf of the Department of Defense.

This letter is to remind you that the focus group will be held on DATE from TIME1 until TIME2 at PLACE in ROOM. Enclosed please find a map with the address and phone number for the hotel, as well as general directions on how to get to the hotel. FOOD TYPE will be provided. To compensate you for your time, you will receive INCENTIVE.

The focus group will include USFHP participants and RAND researchers. During this meeting, you will have the opportunity to share your experiences with and thoughts about USFHP's open enrollment policy, your Military Health System benefits, and your health insurance needs, in a casual environment and with complete confidentiality. Please be assured that RAND will not release any information that can be linked to you.

If you have any questions or are unable to attend, please give us a call toll free at <1-800-xxx-xxxx>.

We look forward to your participation in this important discussion.

Sincerely,

Michael Schoenbaum, PhD　　　　C. Ross Anthony, PhD

H. Sample USFHP Focus Group Confirmation Letter, Eligibles

DATE, 2000

Dear <Mr./Ms.> LASTNAME:

Thank you for agreeing to participate in a focus group discussion of the Uniformed Services Family Health Plan (USFHP), your Military Health System benefits, and your health insurance needs. RAND, an independent non-profit research organization, is conducting this focus group on behalf of the Department of Defense.

This letter is to remind you that the focus group will be held on DATE from TIME1 until TIME2 at PLACE in ROOM. Enclosed please find a map with the address and phone number for the hotel, as well as general directions on how to get to the hotel. FOOD TYPE will be provided. To compensate you for your time, you will receive INCENTIVE.

The focus group will include USFHP participants and RAND researchers. During this meeting, you will have the opportunity to share your experiences with and thoughts about USFHP's open enrollment policy, your Military Health System benefits, and your health insurance needs, in a casual environment and with complete confidentiality. Please be assured that RAND will not release any information that can be linked to you.

If you have any questions or are unable to attend, please give us a call toll free at <1-800-xxx-xxxx>.

We look forward to your participation in this important discussion.

Sincerely,

Michael Schoenbaum, PhD C. Ross Anthony, PhD

I. USFHP Enrolled Focus Group Interview Guide

Good morning. I am X, the moderator of today's focus group. Thank you for coming. I am with RAND in Washington D.C. RAND is a non-profit research institution. I also have XX here with me today and she will be participating in our discussion.

We would like to talk about your experiences so far in the USFHP Program. This discussion is part of a larger study of the USFHP Program undertaken by RAND on behalf of the Department of Defense. The Department of Defense is seeking to better understand how to improve health care services to Military retirees like yourselves.

This discussion is confidential—we won't be associating your names with what you say here—which means I would like everyone to use first names only today. Because this discussion is confidential, I ask that during our discussion you not use specific names of individuals, and this includes your doctor, administrators of your health plan, or any other person. When the discussion is over, please respect the privacy of your fellow group members and do not repeat comments others make during our discussion to anyone outside of this group.

We are taping this discussion today so we don't have to take notes. Does anyone have any objection to this taping?

Only people working on this project will ever hear any of the recordings or read the notes we take. Your participation is voluntary and confidential, and you may refuse to comment on any question that is asked. Nothing you say about a particular facility will ever be made public or reported in any way that will allow you to be identified. Your participation today will not affect the care that you and your family receive. So feel free to say whatever is on your mind.

Before we begin I want to emphasize that you are the experts here today. The reason we are here today is to better understand your experiences in USFHP. There are no right or wrong answers. We want to hear what you think. I'm not planning on doing most of the talking. I do want to make sure that we cover a number of topics in a limited amount of time, so I'll try to keep things moving. There is no need to raise hands. Speak right up. But please respect others when they are talking.

This discussion may last up to about 90 minutes. Is there anyone who can't stay? Before we begin, are there any questions about how we will be conducting this discussion?

Now let's begin.

For nonenrollees: Today's discussion is part of our research into some details of the Uniformed Services Family Health Plan, or USFHP. I understand that you are not enrolled in this plan; we are conducting other discussions with people who are. USFHP is a health insurance option within the Military Health System. It is currently available in seven areas of the country, including where you live. The plan is offered to military retirees and their dependents - even those over 65. USFHP covers a full range of medical services, including prescription drugs.

For beneficiaries under age 65: USFHP is an alternative to TRICARE Prime. For beneficiaries aged 65 and older, USFHP is an alternative to Medicare.

1. Let's start by going around the room and introducing yourself. Please tell us two things about yourself: (1) Your first name, and (2) how long you have been enrolled in the USFHP.
2. What type of health insurance coverage did you have before enrolling in USFHP?

   Civilian employer

   Medigap

   Other TRICARE product.

3. What factors influenced your decision to enroll in USFHP?

   Cost

   Coverage

   Drug benefit

   Choice of doctors

   Location

   Quality

   Recommended by friends/family

   Recommended by health care provider

   Desire for military-sponsored health care

   Only option.

What was the most important factor?

4. How satisfied have you been so far with USFHP?

   What are the plusses? minuses?

   How does this compare to other plans you have been enrolled in?

5. In your opinion, why don't more military retirees like yourselves enroll in the USFHP plan?

6. In the past, retirees such as yourselves could enroll in USFHP only once a year during a month-long enrollment period. Now retirees can enroll in the program at any time. Did this month-long enrollment period once a year affect your enrolling in USFHP?

   Did this month-long enrollment period once a year affect your enrolling in USFHP?

   Did it keep you from getting the care you feel you needed? did it make the care you did receive more expensive?

   How do you think it might affect others?

7. Questions related to Defense Appropriations Bill:

   How would the availability of a low cost prescription drug benefit available through the military health care system affect the interest of you or people like you in USFHP?

   How would the availability of a fee-for-service medigap plan available through the military health care system affect the interest of you or people like you in USFHP?

   How would the availability of increased access to care at MAMC affect the interest of you or people like you in USFHP?

8. In summing up: we will be conducting several of these focus groups in order to gather valuable information from people like you. I'd like to ask for your help: is there anything I haven't asked you that I should have?

Thank you very much for helping us out today. Your feedback will be very useful to us as we try to help DoD understand the health care needs of military retirees. It is all right to talk to others about what we discussed here today, but please remember to respect each other's privacy, and don't mention anyone's name outside this room.

If we have any additional questions or need clarification on any point that was made today, may we contact you?

Would you like to receive a copy of the final report?

If you would like more information about the study, or if you would like to discuss any of these issues further, please don't hesitate to contact me at RAND: (800) xxx xxxx.

J. Maps of Demonstration Areas

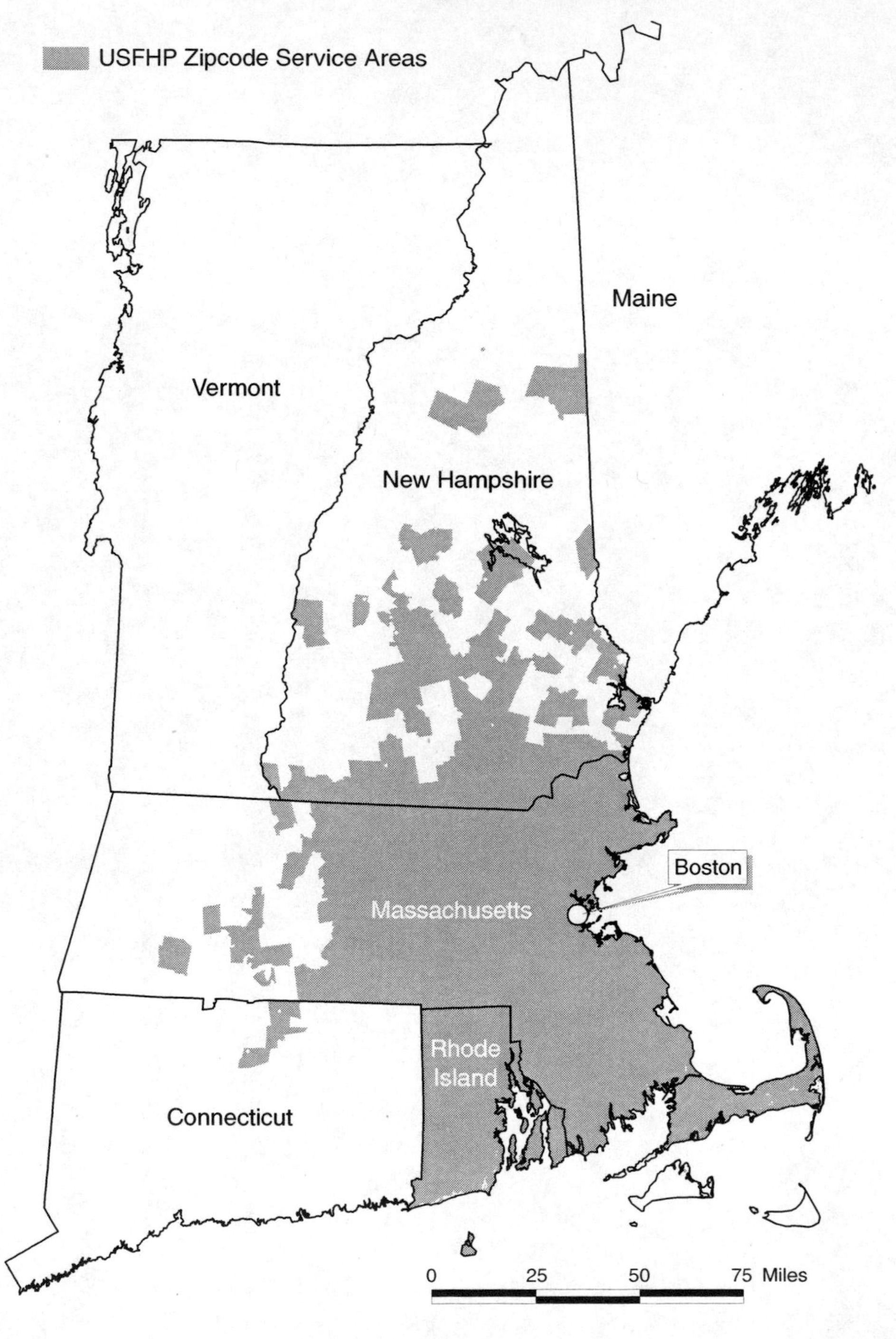
RANDMR1352-M2
USFHP at Brighton Marine
USFHP Zipcode Service Areas
Maine
Vermont
New Hampshire
Boston
Massachusetts
Rhode Island
Connecticut
0
25
50
75 Miles

RANDMR1352-M1

USFHP at Brighton Marine

USFHP Expansion Areas
USFHP Zipcode Service Areas

Maine
Vermont
New Hampshire
Boston
Massachusetts
Rhode Island
Connecticut

0 25 50 75 Miles

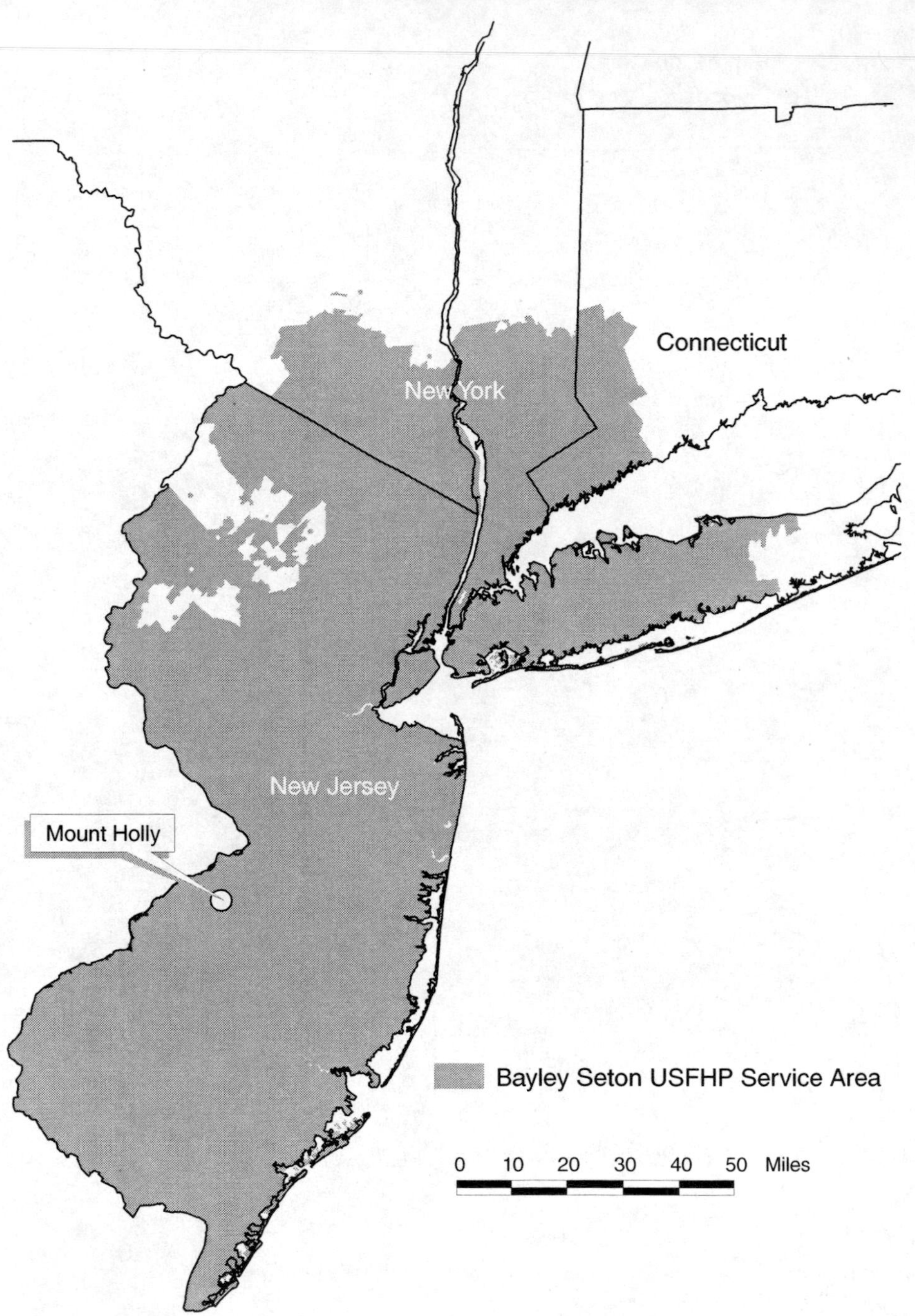
RANDMR1352-M3
Bayley Seton USFHP Service Area
Connecticut
New York
New Jersey
Mount Holly
Bayley Seton USFHP Service Area
0 10 20 30 40 50 Miles

RANDMR1352-M4

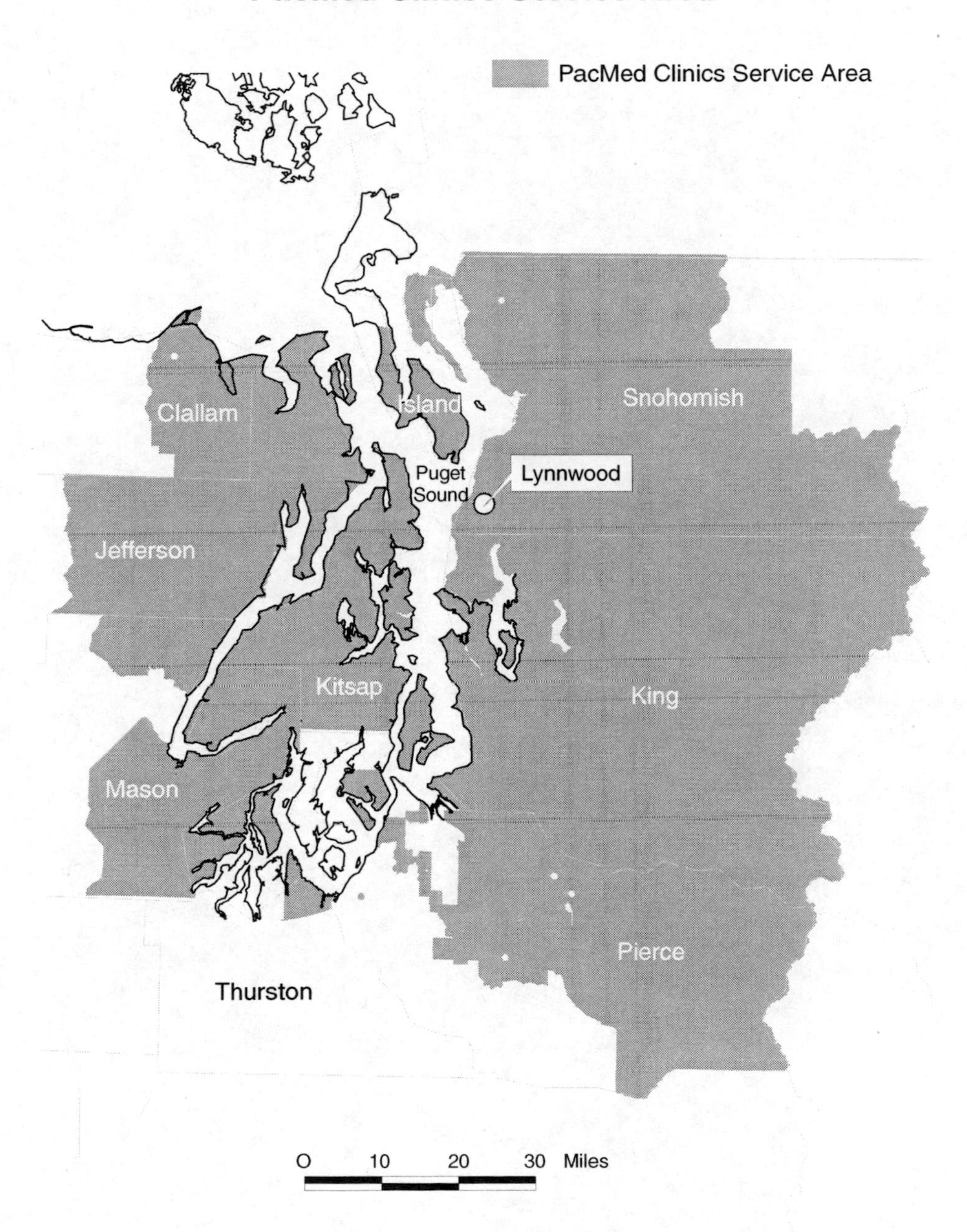